Executive Document, No. 13.]

REPORT

OF PROFESSOR EMMONS,

ON HIS

GEOLOGICAL SURVEY

OF

NORTH CAROLINA.

RALEIGH:
SEATON GALES, PRINTER TO THE LEGISLATURE.

1852.

[Message and Report communicated to the House of Commons. Transmitted to the Senate with a proposition to print 3000 copies. Ordered accordingly.

To the Honorable the General Assembly of the State of North Carolina:

I herewith transmit the Report of Prof. E. EMMONS, who was appointed, under the Act of the last Session, to make a Geological, Mineralogical, and Agricultural Survey of the State.

DAVID S. REID.

EXECUTITE DEPARTMENT,
Raleigh, November 22d, 1852.

REPORT.

To His Excellency, DAVID S. REID,
Governor of North Carolina:

SIR:—

§ 1. Agreeably to the requirement of the Act of the Legislature, passed the 24th January, 1851, authorising a Geological Survey of the State, I herewith present the First Annual Report.

In the discharge of this duty, I have deemed it advisable, at this time, and at this stage of the work, to confine my communication to two principal subjects:—the Soils and Agriculture of the Lower Counties; and the Coal Fields of Rockingham, Stokes, Chatham, and Moore Counties:—the two former occupying the Northern, and the latter the Central, portions of the State. I deem it, however, relevant to the subjects before me, to introduce the statement of such principles of Agriculture and Geology as may be required for the better understanding of these departments; or which are suggested by, or flow immediately from, facts which I have observed.

I hope this course will be approved of, as the subjects are beginning to excite public attention, and are probably among the most important matters to which the public attention as been turned for many years.

§ 2. Among many points which have been established, of late years, there are four of very great importance, namely : *That soils must contain a sufficiency of certain inorganic elements ; that these elements are necessary to the life of the plant ; that no seed can be perfected without them ; and, finally, that they are essential to the life of the animal subsisting on vegetable food.* It follows, from these established points, that some, at least, of the important products of life are derived from the soil ; it being possible to trace them back from the animal, through the plant, to thesoil. From this, it also follows, that the true method of determining the important elements of a soil, is, to analyse the products of life as found in the plant and animal. That which is constantly found in those products, and which can be traced to no other source than the soil, must, of necessity, be regarded as the essential elements of the soil. We can arrive at no other conclusion . and furthermore, by no other method can we reach a correct conclusion. This method has been followed ; and it has resulted in the discovery that the following substances are the essential ones which have been alluded to, namely :—Phosphoric acid, Sulphur, Potash, Soda, Lime, Magnesia, Oxyde of Iron, Silica, Nitrogen, Oxygen, Hydrogen, Carbon, Water, Ammonia, Chlorine, and small quantities of Fluorine.

I might have left out of this list oxygen, hydrogen, nitrogen, carbon, and ammonia, inasmuch as there are, it is supposed, other sources of supply than that of the soil. It is not, however, fully established that, in the arrangements of nature, there is a full provision for this supply, when soils are subjected to high culture, and are required to produce more than five times the amount which they produce spontaneously. It is undoubtedly true, that the vegetable kingdom can sustain itself by the instrumentality of the common sources of supply ; yet, when a species of this kingdom, as Corn, for example, is required to yield its sixty bushels to the acre, we can see no provision for this result in its wild and uncultivated state.

Hence, under culture, where the soil is thus heavily taxed to meet the demands of civilized life, there it fails, in process of time, to supply them, and means to supply them are called for, and even required, in order to sustain the soil under its increased products. Those elements, then, which constitute parts of the atmosphere, as carbonic acid and ammonia, which are furnished in sufficient quantities to plants, growing spontaneously, are not supplied, as I have already hinted, when soils are put under heavy culture; or, they may not be supplied in sufficient quantities to meet the demands of a succession of crops.

§ 3. As I shall have frequent occasion to refer to those elements of soil, generally known as inorganic elements, I propose, in this place, to speak of them; and to state some of their properties, uses, and the sources whence they are derived.

1. Phosphoric Acid.—In its separate state, insulated from other bodies, it is an exceeding sour substance. It is solid, and resembles flakes of snow, when freshly prepared; but, in consequence of its avidity for water, it soon becomes, in the atmosphere, a limpid fluid. Like other sour or acid bodies, it readily combines with potash, soda, lime, and many other bodies; and forms, with them, new compounds, which are called *phosphates*. Hence we have phosphate of lime,—which exists both as a natural substance in rocks, and in the animal kingdom in bones; and it is mainly in this form or combination that it is known in the animal and vegetable kingdom. Animal bodies all contain more or less of phosphate of lime, and probably it is among the most important. The bones, however, contain more of it than other parts; and, in its absence, and when it is diminished in quantity, they are soft and flexible, and unfitted to sustain the weight of the body.

The source of phosphate of lime is the mineral kingdom. Probably all the rocks contain it, sometimes in large masses; but, usually, it seems to be diffused through them in fine particles. When they decompose and disintegrate, it is, of course, mixed with the soil. It is more abundant in granite, greenstone, trap dykes, and volcanic products, than in othe rocks. This fact seems to show that igneous rocks, the *pyro chrystalline*, are the true sources of this important substance. In New York, I discovered, in 1837, a vein of it subordinate to a trap dyke, and in connexion also with the primary or the pyrochrystalline limestone—all of which may be classed together as igneous products. This vein, in one place, was seven feet wide; and hence furnished a large amount of lime for agricultural purposes. In the same section, I discovered the same substance associated with the magnetic oxyde of iron, in small clustered crystals, forming, in many places, more than one-half of the mass.

In the trap dyke, this mineral phosphate is green, less hard than feldspar; and, in the iron bed, in reddish, six-sided prisms. I am careful to mention these facts; for this important substance may exist in the pyrocrystalline rocks of North Carolina; and, if in quantity, would be of great value to the agricultural interests of the State. It should be sought for in rocks of igneous origin; and especially where the magnetic ores of iron exist. In this state it occurs in the marl beds, at the bottom of the shell marl; though it is also diffused through the beds, in masses of small size;—it is in dark rounded and sometimes spiral masses. In this form, it is known under the name of *coprolite*. It is the excrement of marine animals, and contains rather more than 50 per cent. of phosphate of lime; or about the same proportion as it exists in bone. These coprolites are very valuable, and I have indicated the place where they abound the most. They seem to have been collected at the bottom of the shell marl, and to have been subjected to attrition by the waves of the sea. They are associated with quartz pebbles, and generally are black or brown, and almost as hard as quartz. Coprolites are found also in the coal stratas of Rockingham, Stokes, Chatham, and Moore. They have a similar origin to those of the marl beds; but they do not exist in sufficient abundance, as I have seen in either formation, to warrant the expense of extracting them. Still the facts are important, and should not be forgotten. The importance of this substance cannot be doubted, when it is known that Indian corn, wheat, rye, oats, barley, all the cereals, potatoes, and all the tubers and tap-rooted plants contain it, and especially the cereals. A soil destitute of it is totally barren.

2. Sulphur.—It is a substance too well known to require a description. It is not so well known, however, that it is an important element in the vegetable and animal kingdoms. It is found in the auimal tissues. Peas and clover belong to a family of plants in which it is always found. It is an element of oil of vitriol, end hence one of the elements of gypsum. A class of minerals called sulphurets also contain it.

3. Potash.—Equally well known is potash. It is a constituent of granite, existing in the feldspar of the rock in the proportion of 16 per cent. Owing to its presence, this mineral is subject to decomposition, and then forms kaolin; a substance employed in porcelain. Other minerals contain it. The green sand, one of the varieties of marl in the Eastern Counties, owes its fertilizing properties to potash. It plays an important part, both in the organic and inorganic world. It is instrumental in giving solubility to silex; and hence prepares this substance to be taken up into the tissues of plants. In combination with potash, silex is taken up, and made a part of the straw of wheat, rye, and oats, and imparts that strength which is necessary to enable the plant to stand up. When deficient in silex, wheat, rye, and other grains fall, and are injured or destroyed.

Potash is an expensive fertilizer, ranking, in this respect, with phosphate of lime. To this, more than any other element, ashes owe their value as fertilizers. Argilaceous soils contain it, in combination with silex; but the combination is insoluble, and requires the addition of lime to free it from a portion of silica, in order to bring it into a soluble condition.

4. Soda.—This alkali is more abundant than potash, and beds of nitrate of soda exist in climates where no rain falls. Its office, in plants and animals, is not very dissimilar to that of potash. The great source of it is the sea, and beds of rock salt.

5. LIME.—Lime, in some form or combination, is found in all parts of plants and animals. The bark of trees abounds in it, where it serves to protect the wood from injury ; and the testaceous covering of shell-fish, oysters, and clams, together with the integuments of crabs, and the substance of corals. In bones, it is in combination with phosphoric acid ; while, in the bark of plants, it is probably in combination with organic acids ; and, in the testaceous covering, it is combined with carbonic acid. It is more generally diffused in the mineral world than either of the alkalies, or alkaline earths ; and still it is one of those substances which is wanting in soils, and that, too, when they exist in the vicinity of limestone rocks.

Lime also exists in the bones of man, in the proportion of about twelve per cent. From its universal diffusion in the vegetable and animal kingdoms, it is evident it is one of the important elements of the soil. Its presence in the soil, however, is not so common as might be expected from its great abundance in the mineral kingdom. We have no calcareous soils, even upon our limestone, though there may be patches of a few yards in extent where lime is the principal substance. Much uncertainty prevails in the use of lime. I shall, however, reserve what I have to say, under this head, until I have occasion to speak of the use of marls.

6. MAGNESIA.—It is often maintained that magnesia is hurtful to soils. When caustic, it does not become mild so soon as lime ; and hence is liable to absorb the water which is required by plants. Yet the phosphate of magnesia is a constant element of the wheat, rye, and corn, as well also as in all vegetable food. It exists in soil; is important to fertility ; but is less so than lime. Its source is in the magnesian rocks, as they are termed, such as soapstone and steatite talcose slate. It is far less less soluble than lime. When barrenness appears in connexion with serpentine, it is not because magnesia is injurious, but because other earths are wanting or absent. Magnesian limestones have rarely, if ever, injured vegetation in this country, though prejudices exist abroad.

7. Silica.—In flint and rock crystals, we have examples of this earth. It is harder than glass ; and, in the common form in which it is found, is insoluble. It appears, in this form, to be one of the most indifferent of all bodies. Yet it is found in such combinations that it is freely taken up by the roots of grasses, and plants of this family. It constitutes a very large proportion of the earth's crust. In soils, it varies in quantity. Its proper proportion is about 85 per cent, ; but good soils often contain less, and two or three per cent. more. The office which silex performs is to preserve a due amount of the coarser matter : for, when a soil is composed of impalpable matter, it is comparatively barren ; and silex, from its excessive hardness, resists the wear and action of the elements. By being commingled with clay, it imparts porosity and looseness ; permits the roots to penetrate deeply ; while, at the same time, air and moisture permeate through the mass as far as roots can find their way. The use of silex is, therefore, partly mechanical and partly physiological ; being necessary in soils to preserve porosity, and particularly necessary to the cereals to protect the straw and kernels, and give elasticity and strength to the whole plant. So abundant is this substance, however, that it is never necessary to add it to the soil, except for mechanical use ;—potash and lime are often added for the purpose of freeing it from its insoluble combinations, when the grains are special objects of culture.

8. Oxide of Iron.—Analyses of organic bodies prove the existence of iron in them ; and in those animals which have red blood, it is satisfactorily demonstrated that it serves to maintain the heat of the body. In addition to this, its salutary effects upon the human system prove, also, that it performs some other office besides. It is, then, an important element, physiologically. In the soil, however, it is supposed to be concerned in developing or forming ammonia. Iron, by itself, is rarely used as a fertilizer. The oxide, however, taken from the smith's forge, intermixed with refuse matter collecting about a smith's shop, is often highly beneficial to fruit trees ; and pear trees, especially, have derived essential benefit by the application.

9. Ammonia.—It is known by everybody, under the name of hartshorn. In this State, however, it is not applied to plants, or to soils. It belongs both to the soil and atmosphere. Nitrogen, an element of the atmosphere and ammonia, is an essential constituent of the cereals, being one of the components of gluten, or the pasty part of flour. It is supposed that the salts of ammonia are the media through which nitrogen gains access to the grain; and there are results of experiments which go to prove that, if it is wished to increase the wheat crop, it requires the addition of substances which will furnish ammonia—that the fertilizers I have noticed in the foregoing paragraphs are not sufficient, or are inefficient, in the case of wheat, unless they are mingled with ammoniacal compounds. Phosphate of lime will greatly increase the turnip crop; yet, if applied to wheat, when the soil is unexhausted, with a view to increase the product, it fails, unless ammonia is furnished also. Yet, on poor and exhausted lands, phosphate of lime is known to produce surprising effects. In fresh soils, plaster and charcoal readily absorb ammonia.

Water.—This element is the great solvent of the different inorganic bodies upon which I have been speaking. Nothing can act and become beneficial to vegetation, until it is dissolved in water. It is, therefore, the medium through which all the essential substances find their way into animal and vegetable tissues. The effect of much is injurious. Standing, as it frequently does upon soils, it diminishes their temperature, and maintains them in a state permanently too low to admit of the cultivation of the valuable plants. Draining lands of superfluous water is, in effect, raising their temperature several degrees. Where there is too much water, another condition exists incompatible with that cultivation which the cereals require;—the soil is too compact, and it cannot be made porous while in that condition. Draining, therefore, makes the soil warm and loose;—conditions essential to the growth of the most important productions.

Water is indispensable, and no seed can germinate without it. Soils differ as to the force with which they retain or absorb it.

Silicious soils part with it readily, and absorb it slowly. Of all substances which retain water, finely divided peaty matter is the strongest ;—it exceeds clay and marl. Next to peaty marl, fine marl, containing some organic matter, ranks the next.

10. CARBONIC ACID.—The atmosphere is regarded as the source from which it is obtained by plants. In this combination, it is always produced, and it is generated in the soil. Carbonic acid is a solvent. Water charged with it dissolves rocks. The almost insoluble phosphate of lime is thus dissolved in water by its aid.

Leaves are supposed to absorb it from the atmosphere ; and to obtain, in this way, the carbon required to build up their structure. Still, the water in the soil holds it in solution ; and it is, under those circumstances, furnished the plant by its roots. This seems to be the channel through which carbonic acid may more naturally course through its tissues, when it is assimilated. Carbonaceous, or peaty matters, also supply it.

11. CHLORINE.—Common salt is a combination of soda and this substance. The term *chloride* is applied to such combinations. By itself it is a poison ; in combination with soda, it is a fertilizer. Its true value, however, is not well settled. Some esteem it highly ; others do not. On wheat its effects are scarcely perceptible. It promotes the growth and yield of plums, and it may be taken up in sufficient quantity to give them a saline taste.

12. FLUORINE.—It is found in combination with lime, constituting the mineral called *fluor spar*. Although it is a rare substance, yet it is found in the enamel of teeth, in bones, and in milk. It is an associate of phosphate of lime. It is never added to soils by itself ; but, as it accompanies phosphate of lime, I believe, in all cases, it is applied when phosphate of lime is used.*

* In the foregoing pages, I have used the word *element* in a different sense from that in which it is employed in chemical works. Real elements, or simple bodies, are never employed as fertilizers. They must be compounded, before they are received as elements.

The foregoing substances are detected in the ash of plants. When plants, wood, coal, &c. are burned, the ash that remains is called the inorganic part of the plant. If they are burnt carefully, in a proper flame, we find that the particles composing the inorganic part preserves a reticulated structure, and often appears as woven. It seems, therefore, to be designed to perform the part of a skeleton to the plant, and give it firmness and elasticity. Even the delicate petal of a plant has its fine woven skeleton. Two things should be observed of the inorganic part:—plants differ among themselves in the quantity they contain; and the parts of the same plant differ also in this respect. These two facts lie at the foundation of an improved and refined system of agriculture. That improved system would consist in adapting the quantity of the inorganic elements to the special wants of the plant. In the present state of our knowledge, something can be done, but it must be done imperfectly. Hundreds of acres under cultivation can receive only rough and imperfect tillage. Special agriculture,—that which is conducted to meet the special wants of the crop,—must limit and confine the operations to small plantations. This special agriculture is, in part, observed, when the planter and farmer puts his wheat on soils best adapted to wheat; or when his Indian corn and potatoes are cultivated on soil best adapted to them. But when the system of artificial farming is undertaken, it is necessary that more knowledge of the soil should be obtained than can be procured by simple inspection. A full knowledge of the composition of soils is the first step towards real improvement in the right direction. To carry the improvement to perfection, a full knowledge of the composition of plants is also necessary But plants vary much in their relations to light and shade, heat and cold, to dry and wet soils. The conditions of vegetation best adapted to plants, or to crops, must receive study;—the reasons why they vary should be determined. All these points require a knowledge of the economy of vegetation, or of its physiology. The range of knowledge required for the practice of a special agriculture, or where special adaptations are attempted, is by no means confined to a small compass. This is not to be regarded as a discouraging feature. To progress, however, requires the utmost patience, and the great danger is, that discouragements will spring up at the slow progress which is made, and a right road will be abandoned, because it can be seen only for short distances ; the end is yet hid in mist and doubt.

The organic part of a plant, is that which is consumed during combustion : the products being volatile, are all dissipated. It forms by far the greatest part of the vegetable. Its source is regarded, by Liebig, as the atmosphere. Still, the soil no doubt furnishes it, in the form of organic salts, which are known under the names of crenic and apocrenic acids in combination with alkaline matters. These are derived from humus originally, plants first changing in water into peat, which, in itself, is scarcely soluble, but which becomes so, in part, by the action of lime.

4. It has been stated that certain elements are essential to a productive soil. Knowing before hand what those elements are, it seems to be plain what course our enquiries should take when directed to the improvement of any given soil. If a a crop is defective in quality, and falls short of its former yield, it is evident thas there is a want of those elements which have just been described ; the course to be taken then, is to analyze the soil, especially those patches where the failure in quantity is the greatest. If it is found deficient, in some of those elements, we are put in a way to correct the evil. It is possible that this report may fall into the hands of a few who may wish to know the chemical composition of soil, really poor, as well as those which are rich, I shall proceed to give the results of an analysis of several kinds of soil, in order to make my readers better acquainted with their composition, that they may be used for a comparison hereafter ; and as they are taken from well known plantations in different parts of the State, the results may also be regarded with more interest than if selected from books. The planter wants to know the reason of the failure of his soil to produce its customary crops. But til-

lage must be taken into the account, as well as the season; and, indeed, all those variations in seasons, time of planting, favorable conditions of soil, &c. which are necessary to arrive at a true conclusion. If the failures belong to successive seasons, equally favorable to the crop, there can be no doubt they arise from a deficiency of one or more of the inorganic elements of the soil. There may be a kind of *a priori* determination of the cause of failures by looking back for several years, and calling to mind the kind of crops taken from it. If they have been cereal, then it is highly probable that the phosphates and potash have been deficient. If tobacco has been cultivated several successive seasons, much potash has been carried away; or, if the potash is deficient in the soil originally, lime may take its place.

It is proper to say, in this place, that a registry of crops will be found useful; for, if the quantity removed from the soil is noted, the number of pounds removed of either potash, phosphate of lime, or lime, may be calculated.

§ 5. I shall proceed now to give a number of analyses of soils, for the double purpose I have stated. The first class belong to the poorer soils, and have their representatives too common upon the Atlantic slope of this State. It would be unjust to attribute the faults of the soil entirely to culture and bad husbandry. The truth is, they have a sandy basis, and when cultivated for several years in succession, without returning something in the way of fertilizers, a fine marine sand is exposed, which, in some plaees, is so loose in its texture as to permit the vegetable matter to be blown from its surface.

The first examples of soil show a deficiency in several elements; and from these deficiencies others have followed, which affect it mechanically.

The samples were taken from an elevated part of the Panola plantation at Tarboro'. The proprietors are R. Dancy and Norfleet, who are pursuing agriculture as a profession. Their success is proof that they are pursuing the right road.

	SURFACE.	SUBSOIL.
Water,	1.13	0.92
Organic matter,	2.93	1.72
Silex,	93.70	94.40
Alumina and peroxyde of Iiron,	1.33	2.40
Carbonate of lime,	.28	.20
Phosphate of lime, inappreciable in 100 grains,	.00	.00
Magnesia,	.14	.06
Soluble silica,	.11	.09
Potash,	.03	.04
Soda,	.01	.03

The palpable deficiencies, both in the surface and subsoil, are :—1. Alumina and per oxyde of iron ; 2. Lime ; 3. Phosphate of lime ; 4. Potash and soda ; 5. Organic matter.

These deficieneies leave a great excess of fine sand. It is proper to remark, that the samples do not represent the character of the field, but simply an acre or two which overlooks it. It is a sandy knoll, and its soil is one of those extremes which are often met with. Such eminences suffer more by cultivation than the lower parts , and, hence, are liable to become bare, unless more care is bestowed upon their cultivation.

The system formerly pursued, or until the plantation was purchased by its present proprietors, was the common one ;—a system which, if carried out, would end in total barrenness :—for that is the tendency when fields are cultivated until they fall off greatly in their products, when a period of rest is allowed for their restoration. The rest being the only mode by which its restoration is expected,—for, if it is true that certain elements are necessary to fertility, and these exist only in the soil, and they are removed in the crops, then, rest cannot restore them,—the only effect of rest is to give time for the growth of trees, whose roots penetiate deeply, and bring up from greater depths than the cultivated plants can, these inorganic elements, which, when assimilated and formed into leaves and other parts, either fall again to the ground in due time, and give back to the earth what had been taken from it. By this process, the inorganic matter is transferred from the depth of many feet to the surface. Hence, after years of rest, there is an appearance of renovation.—But, let the cropping re-commence, and the result will be seen; for it will require only half the time to exhaust the soil. It will require, in a shorter time than before, its period of rest. By this course, *or this rest system,* the soil loses more of its fertilizing matter, until, finally, it will be so far exhausted, that only the most obscure plants can find a foothold.

To return to the consideration of the soils. Knowing their deficiency, how can they be corrected?—or what course do the proprietors propose to renovate this field?—seeing that they reject the rest system as a means to this end,—being satisfied that they would only entail a barren lot to posterity. One of the means resorted to was to furnish or supply *ashes.* The plantation bordering upon Tar river supplied decaying logs, brush and leaves. These were burnt, and gave them about 5000 bushels, at a trifling cost. The ashes contain all those elements which have been removed in the crops in former years. It is this fact which makes ashes valuable for this purpose. I notice this fact, not so much that this method was resorted to, in this instance, for a special purpose; but because, in all parts of the State, it is possible to pursue the same plan. The decaying vegetable matter on the borders of plantations is enormous; and while the ashes of refuse matter can be procured at a cost so trifling, it is folly to purchase bones or guano. But the soil is deficient in organic matter. To supply this, various means may be resorted to. Anything which has lime contains it, and will supply. Straw, leaves, chips of the wood-yard, peat, or peaty soils hauled to the yards and trod by cattle, are the cheapest. To carry out a system of supply, for losses of all kinds sustained upon a plantation, the most effectual means will be to give to one one or two laborers the business of collecting ferilizers of all kinds. It is a labor which the disabled and inffrm can perform, especially if aided by a team consisting of a single mule and cart.

By this plan, the capital required to be expended in the purchase of guano and bones is much diminished; for in the ashes we find phosphate of lime, carbonate of lime, iron, potash, soda, and magnesia. In all these substances the soil is deficient. It will produce sassafras and high briars, because their roots penetrate deeply, and find a store of food unconsumed by cultivation. But the cereals, the plants most valuable to man, though they might exist and produce a small crop, yet they cannot pay the planter, in its present state, the cost of tillage. The proprietors are pursuing, therefore, the most judicious and the cheapest plan to make this field again fertile. They have no marl; but, in the ashes of useless logs and brush, they find a better material.

Similar to the foregoing soil is one which I analyzed for Mr. Benjamin Brown, of Pitt County, near Greenville. Its composition is as follows :—

Water,	1.20
Organic matter,	1.30
Silex,	94.75
Alumina and per oxyde of iron,	1.48
Lime,	1.02
Phosphates inappreciable,	.00
Magnesia,	trace.
Potash,	.01
Soda,	.02
Soluble Silica,	.05

This soil is less fertile than that of a portion of the Panola farm. Its color is a light ash, and the sand quite fine, and uniformly so over a large field. The same plan is proper for the latter as the former. I may here suggest that, along with similar means for its restoration, the cheap material of bogs and wet places, clays and unexhausted surface soils, will meet some of the wants of the case. Their use turns however, upon the value of labor, or the cost of transportation.

Of the same class is a soil obtained at Mr. Swift's, Ringwood, near Halifax. It is an ash gray sandy soil; but the particles are angular and sharp. I found its composition to be similar to that of the two preceding.

Water,	1.30
Organic Matter,	1.35
Silex,	94.15
Alumina and per oxyde of Iron,	1.80
Lime,	.15
Magnesia,	.01
Potash,	.01
Soda,	.01

So, also, I may add to the foregoing a soil taken from the plantation of Mr. Cromarty, near Elizabeth, on the Cape Fear, as it contains :—

Water,	1.250
Organic Matter,	1.500
Silex,	94.800
Alumina and Iron,	.650
Lime,	.010
Magnesia,	trace.
Potash,	.006
Soda,	.004

This soil was not tested for phosphoric acid. There is a remarkable deficiency of alumina and iron, in which the phosphoric acid, or phosphate of lime, would be found, if found at all. It is highly probable that it exists only in an extremely small per centage.

Some portions of the plantation of Mr. Pope, in Halifax, and of Dr. Eppes, have the same excess of sand, and a deficiency of the most important elements, as may be seen by the following statement of their composition:

	Pope Sub-Soil.	Dr. Eppes.
Water,	1.26	1.39
Organic matter,	2.20	2.70

Silex,	92.08	92.56
Alumina and Per Oxide,		
Of Iron	1.64	2.70
Carb. Lime,	.08	.13
Magnesia,	.86	.24
Soluble Silica,	.27	.10
Potash,	.01	traces.
Soda,	.00	.18

The field of Dr. Eppes was in rye.

There is less exhaustion in these two soils, than in some of the foregoing examples. It is evident, however, that there is a great excess of sand, and a deficiency of Alumina and Iron. Their presence is required to form a suitable basis upon which to apply soluble fertilizers. The foregoing examples of soil belong, it will be conceded, to one class. There are, in all of them, both excesses of one element and defects in others. They are strictly poor soils from deficiency, and must be improved by addition of those elements which are evidently, in a measure, absent. While upon this subject, it is quite necessary to add, that, although soils of the composition, which the foregoing possess, are unsuited to the cultivation of the cereals, still another class of vegetables find in such soils sufficient nutriment. Bordering the coast of the State, the surface is undulating, and rises at intervals into rounded eminences, and sinks into slight depressions, with tolerably well defined borders or rims, forming, as a whole, a rolling surface. This border is always formed of a marine sand, which, in the driest parts, is intermixed with vegetable matter, upon the slopes and tops of the eminences, while in the bottom of the bowl, from depressions, the vegetable matter occurs in much greater quantities, and even beds of peat are often found some two or three feet in thickness. Marine sand, however, is the basis of the soil, and when washed and dried, is often pure enough for glass. The tops of the rounded eminences are generally whitened with oyster and clam shells, which are bleaching, and slowly disintegrating. This light sandy soil, which, under careless cultivation, would be converted into blowing sand in a few years, yields large profits in the culture of the ground-pea. This plant, like the common pea and clover, is constiutionally adapted to this soil, and yet the inorganic matter, which is found in the ash of the pea, is by no means inconsiderable.

The plantation of Mr. Nixon is about ten miles east of Wilmington. It is situated immediately upon the coast, with a soil and surface I have just described. This gentleman cultivates the ground pea. The annual value of the crop rarely falls below six thousand dollars. This pea soil has ninety-five per cent. of silica, and from one and a half to two per cent. of organic matter, and less than one per cent of alumina and per oxyde of iron, and about ten hundred per cent of lime. I have not stated, I perceive, that beneath the sands there is a stratum of brick clay: it is sometimes within two feet of the surface, and at others ten feet. It is no doubt a layer which exerts an important influence upon the cultivation of the loose surface sands. It holds water, and hence aids in supplying water to the sands which rest upon it.

The pea husbandry has to be conducted with care. It consists of an alteration of crops and a rest of one year. Thus to the pea crop succeeds rye, oats or millet; the latter is recommended; and then rest for one year. The soil should be disturbed as little as possible; even cattle should not be allowed to roam and feed upon the field, inasmuch as they break up the surface sand. The roller is an admirable instrument for these lands. The pea is planted in hills about two and a half to 3 feet apart. One pea is sufficient for a hill. They are hoed sufficiently to keep out the grass. The yield is from fifty to seventy-five bushels per acre. One hand can cultivate five acres. The ash or inorganic matter of the pea vine is large, amounting to 10.25. The amount of lime, uncombined with phosphoric acid, shows that it is a lime plant. No doubt the comminuted shells constitute an important element in this sandy soil. The lime is of averagable quality.

The decomposing oyster shells on the summits and slopes furnish lime slowly to the soil by the aid of carbonic acid.

In this connection, it will be instructive to exhibit the composition of a good soil—one which is adapted to a general cultivation: one which produces corn, oats, potatoes, and even wheat. I find soils of this description in many places, and have several analyses of them: I shall select from my note book the following. The first is from Mr. Swift's plantation, in Halifax county. Its color is brown, and it is composed as follows:

Water,	4.50
Organic Matter,	5.20
Silex,	74.30
Alumina and per oxyde of iron,	14.00
Phosphoric acid appreciable.	
Carbonate of lime,	.40
Magnesia,	.20
Potash,	.05
Soda,	.03
	98.68

If this soil is compared with the gray sandy soil of the same plantation, the differences are too striking to escape notice. To remark upon each element: The water which is obtained by drying, at 400° of fahrenheit, is the quantity which experience and observation prove to be about right. The organic matter, which is afterwards obtained by heating to redness, is an essential material, from which organic salts, soluble salts of lime, and the alkalies are formed, and which, under the forms of the so called crenates, may be introduced into the tissues of plants. It is in those forms, and in these combinations, that organic matter, at least in part, is introduced into the system of plants. Where organic matter is absent from a soil, experiment proves that ripe seed fail to be produced. Silex is really the basis of all soils, and is an indestructible and almost unchangeable substance: it becomes soluble in the presence of the alkalies, and gives strength to the stalks of the cereals. It is rarely deficient in quantity in soils proper, for it seems to me that deposites of vegetable matter, alone, can scarcely be called soils; still, as certain plants are produced upon such deposits, they must be ranked with soils.

6. Alumina never enters into the composition of plants or of animals, yet it is an essential element of all soils. The function it performs is that of a cement; as it obviates or diminishes the porosity of soils, and prevents the too speedy percolation of water through them. It is often in excess; in many others too little, as in the first examples given of the composition of one class of soils.

The function which iron performs in a soil is not well determined. It is, however, an element of considerable importance to living bodies, and is always present in them. But, it is supposed by some physiologists, that it is instrumental in furnishing nitrogen. It exists in two states—protoxyde and deuotoxyde, and is readily changed from the latter to the former state, in the presence of water and organic matter. Water is decomposed by the first, it is supposed, and the hydrogen of the water is set free, which combines with nitrogen of the air and forms ammonia. There is a mutual action also, between the organic matter and the deutoxyde, by which the deuotoxyde is changed to the protoxyde. In this last change there is a step preparatory to its change into an acid, which it combines with lime and alkalies. These chemical changes are by no means improbable; indeed, they are quite agreeable to what is known in analogous cases.

Phosphate of lime is one of the most essential elements of soils. No living being, whether animal or vegetable, is destitute of it. Bones, muscles, nerve, brain and blood, contain it. Milk would be unsuited to the young or old if it was not rich in it. The quantity of food is always greater than the system can take up. It becomes excrementitious matters, and hence their value as fertilizers.

Lime and Magnesia are known in animal and vegetable structures. Potash is equally essential with the phosphates. Soda is undoubtedly essential also, but perhaps less so than potash ; at least it is not so expensive, and can be supplied at a cheaper rate. Water is a solvent for all bodies which enter into the structure of plants. All matters must exist in solution before they can be received into the structure of plants or animals. There is no growth in the absence of water. It will be observed, that the most important elements exist in the smallest quantities. Hence it is, that they are liable to be removed in many states, and as they are not at all abundant in nature, they are expensive to supply.

7. The two following are examples of good soils, whose elements exist at least in fair proportions, though the organic matter is five times the amount usually present. They were taken from the plantation of Mr. Purnell, of Halifax County. The first is a dark brown soil ; the other a dark peaty soil :— both breaking into angular fragments when dried.

	1. Cultivated Swamp Soil.	2. Uncultivated Swamp Soil.
Water,	4.27	5.29
Organic matter,	12.16	10.30
Silex,	76.64	78.72
Alumina & per oxyde of iron,	5.69	3.78
Lime,	.12	.17
Magnesia,	.6	—
Soluble Silica,	1.12	.36
Potash,	.08	.20

Phosphates appreciable in both samples.

Soils of the foregoing descriptions are durable. Still, in process of time, the relations of the elements are changed ; the organic matter gradually disappears ; and, with that change, the capacity for retaining water is diminished, and the more expensive elements, phosphoric acid, potash, and soda, will have been removed in the crops.

When any of the foregoing soils were submitted to the action of pure water, they furnished soluble salts, consisting of sulphates and chlorides, in the proportion of 2.50 per cent. These salts are present in all soils. Sulphur and chlorine also are essential constituents of organized matter.

A soil from Mr. Bullock's plantation, near Tarboro', is another example of a soil in good condition. It contains

Water,	4.00
Organic matter,	3.20
Alumina and per oxyde of iron,	5.50
Silex,	86.80
Lime,	.12
Magnesia,	.24
Potash,	trace
Phosphoric, not tested.	

The deposits of the Roanoke, in Halifax County, are sometimes used as fertilizers, and they answer very well the end proposed. They may be regarded as having mainly the chemical constitution contained in the following

Water,	5.784
Organic matter,	8·160
Silex,	74.540
Alumina and per oxyde of iron,	10.560
Lime,	.248
Magnesia,	.360
Soluble Silica,	.038
Potash,	.220
Soda,	.100

The Roanoke brings down sediments of great value ; and they may be profitably employed in enriching adjacent fields. It is evident the upper country is wasted by its turbid branches; and it is very desirable that some cheap, practicable means might be devised, by which its sediments may be diminished.

A deposit, or sediment, of the Tar River, upon the Panola plantation, does not differ materially from the Roanoke sediments at Halifax.

In this connection, I will introduce the composition of a good wheat soil; known to be good, by observation and yield of the crop. It is an example taken from the County of Perquimans, and from a field of 30 acres at least. The surface appearance is perfectly uniform, and so was the fine waving crop, as the wind played gently over it.

Water,	4.63
Organic matter,	4.69
Silex,	85.73
Alumina and iron,	7.40
Lime,	.16
Magnesia,	.02
Soluble Silica,	.92
Potash,	.06
Soda,	.01

Phosphoric acid appreciable in 200 grains.

The soil is stiff, or of that character which is denominated argillaceous. This large field was remarkable for the uniformity of the crop, both as to height and advancement; and it is one of the best examples of a good wheat soil in the State—equalling, in every respect, the soils of Genessee and Monroe Counties in New York. This kind of soil, however, would not have been noticed here, if it had been confined to a single field. A large proportion of the county has this excellent wheat soil.

Another example of a good wheat soil is that of Mr. Cromarty's, near Elizabeth, on the Cape Fear River.

Water,	5.20
Organic matter,	9.20
Silex,	74.03

Per oxyde of iron and Alumina,	10.40
Lime,	.40
Magnesia.	.10
Potash,	.03
Soda	trace.

The phosphoric acid remains to be determined.

I proposed to state the composition of a good soil, adapted to general cultivation. The wheat soils which have been added to those thus regarded, will serve as an illustration of the slight difference which is required to convert a good general soil into one adapted to a specific use ; and this leads me to remark, that it is not so much the chemical as the physical constitution which produces the change. For the addition of clay is what makes the difference ; and I have already stated the fact that clay, or alumina, never enters into the constitution of a plant or animal. It is never assimilated in the animal body, though taken into the digestive organs ; and it is never taken up by the roots of plants.

As clay is regarded by some as a fertilizer—and, indeed, there can be no doubt of its good effects when applied to lands deficient in clay,—we may understand the principle upon which it operates. It is, however, true, that it may carry with it potash, and other necessary elements of growth ; still we should be slow to advise its use on that principle ; inasmuch as they are usually in proportions too small to justify the expense of hauling. Circumstances must be very favorable to make such a procedure profitable, except in cases which will come up soon for consideration.

We have seen, in some of the foregoing examples of the composition of soils, that there are certain extremes—I might say anomalies,—in soils : they may consist of sand, with a trifle of organic matter ; and, when thus composed, they will produce something, if the sands are fixed. As there are sandy soils at one extreme of the varieties, so there are peaty soils, or really peats, at another. It has been generally supposed that the latter are unsuited to cultivation, *per se* ; and in fact, this must be regarded as a fixed fact ; yet it is surprising how little inorganic matter is required to convert the peaty soils into highly productive ones. If a seed is cast into a bed of pure peat, a bed purely organic, or even a marl bed, it soon sends forth its blade ; it grows for a short time, and then turns yellow and dies : it is the last of the seed, and it fails to reproduce itself in the midst of a magazine of food. This fact is applicable to the kind of soils of which I am about to speak.

The State of North Carolina owns large tracts of land, in the Eastern Counties bordering the coast, the soil of which is eminently of vegetable origin. The tract which I am to notice is situated in the County of Carteret, and is known as the Open Ground Prairie. It is within six or eight miles, North, of the town of Beaufort, and contains over 80,000 acres. In form it is an oblong, being about twice as long as broad, its longer axis extending E. N. E. On the S. E. side, several creeks penetrate into it ; or, rather, Core Sound sends into it three short arms, which are known as creeks. The 1st is Ward's Creek ; 2. Willis Creek ; 3. Oyster creek. At the extreme S. E. border is the North river ; opposite to which, on the other side, is Adams Creek, and then South River. The creeks and rivers are channels which render access to the Open Grounds more feasible ; for it should be known that nature has fortified these grounds by thickets of brambles.

Around, and upon the outskirts of the Open Grounds, are ridges, which seem to have enclosed, at no very distant day, a body of water, which was probably shallow and fresh, and communicated with the Sound. By the special direction of His Excellency, the Governor, I visited this tract, mainly for the purpose of determining how far it is susceptible of reclamation and cultivation. In this enterprise I was aided very efficiently, and indeed kindly, by gentlemen residing at Beaufort ; especially by Dr. Arendell, Mr. Hellen, and Capt. Farrar.

We gained access to the Prairie through Ward's Creek, a branch of the North River. In our course we passed over

Piney Ridge, which has a breadth of one-fourth of a mile, and which has vegetable mould to the depth of $3\frac{1}{2}$ feet, which reposes upon clay. Beyond this ridge is a zone called the Everglades, which are productive and rich. The soil is between 5 and 6 feet, and reposes upon a sandy clay, and in which there lies buried an ancient forest. Sounding at various points as we passed over the Everglades, we found bottom at various depths, varying from 3 to 10 feet, and frequently the sound penetrated prostrate logs of considerable size. The soundings in the Open Prairie were sufficiently numerous to prove a great uniformity in the covering of an ancient sea bottom, which, no doubt, was finally changed into a fresh water lake or sound, which, by the progress of slight elevations, from time to time, raised the bottom above high water mark. In consequence of these changes, vegetables peculiar to marshy districts sprang up. These vegetables, especially the humble kinds, belong to the family of mosses, but consist mainly of a sphagnum. Trees also grew upon this bottom, particularly the bay; and they evidently attained a large size. But the moss growing luxuriantly, has finally raised the surface in some places 10 feet upon the sand. This renders the present surface less suitable for the growth of trees. The Open Ground Prairie presents a level surface, so far as can be seen, covered with an humble vegetation of sedge and moss, with here and there a solitary pine. Surrounding, however, this extensive field, there are lofty pines, interspersed with oaks and chincopin. The soundings prove a field far more productive in trees than the present. This fact, taken by itself, looks favorably towards the soil which bore them. In general, however, they were probably bays, and these are not indicative of good soil. The soil when brought up from the lowest points we could reach, was, to the eye, composed of vegetable matter, with some sandy soil. It was found loose towards the surface, at least to the depth of 18 or 20 inches. It is even a sponge, and holds and retains water like a sponge. The train of argument which I was disposed to adopt was, if this soil has been competent to produce trees and perfect seed, capable of reproducing species, it may be put under successful culture, provided it is put into a favorable physical condition. Still this argument does not hold good. I was inclined to adopt it before I knew the exact chemical composition of the soil. It is fully established, as I have had occasion to say, that a soil, purely vegetable, will not produce the cereals; neither will one of pure sand or pure lime produce them. Organic matter is an essential constituent, as is sand and clay and lime, and yet neither by itself can produce a ripe seed. There must be a mixture of organic and inorganic matter. The latter must also consist of several elements. These are important principles which may be applied to the question which was to be settled by this examination. The characterstics of the vegetable material were found uniform, and hence it did not require numerous analyses for to determine the nature of the soil. Hence, only four were taken for this purpose.

These samples were taken from a depth of eighteen or twenty inches. The chemical examination of the specimens taken, resulted, uniformly, in this: One hundred grains gave three per cent. only of inorganic matter, proving the almost total absence of any earthy compound—the three per cent. consisting of the ash of the vegetable matter. This ash contained silica, phosphate of lime and peroxyde of iron, lime, magnesia and potash, or the same elements which are usually found in the ash of plants belonging to marshes. A soil thus constituted is not susceptible of a profitable cultivation; certainly not competent to produce corn and other cereals, unless it be rice. The question then comes up, is there a remedy? Can the Open Ground Prairie be brought into a condition to warrant an attempt at cultivation, for a reasonable expense? It is plain enough, that the first step is to lay the prairie or portions of it dry, by draining. This is feasible, as it is proved by a competent engineer, to be sixteen feet above high water, and to be also above the storm tides of the coast. The first effect of draining will be to reduce the level of the prairie about eighteen inches. The surface, and, indeed, the whole bod , will becrme more compact and close, and the unchanged organic matter will, in time, decompose. Another effect will be, to raise the temperature of the soil, which is now constantly below that of the dry or drained fields. Both of these results will promote a vegetation of better kind; but they cannot change materially the composition of the soil. There is still something more which must be done. The method which has hitherto been pursued with soils of this kind, is, to add quick lime only, under the impression that it promotes the speedy decomposition of the vegetable matter, and converts it into an element for plants. This practice, however, is not founded upon just views: it is at least defective, and besides, it is too expensive. The trials, too, of this method, have failed; not because lime is injurious. It does not go far enough to add lime, as this is a single element. Now, it is proved, I believe, that these soils are unproductive, for the want of inorganic matter; or, in other words, because the earthy bodies are absent; not because lime is absent more than the other earth, but because they are all absent. It follows, then, if the foregoing principles are true, that what is required, is the addition of soil. Take any of the uncultivated soils, marsh mud and sand, anything of the kind at hand, which consists of earthy matter, and apply it as a dressing. Experience proves that the quantity required is not large; that what planters call a heavy dressing, is sufficient. In using soil, instead of lime, as a fertilizer, there is added to the vegetable mould, the elements necessary for the production of the higher order of plants, the grains or cereals. Lime, magnesia, phosphoric acid, alumina and iron, potash, soda, etc., are all incorporated with the organic matter, which, together, constitute a good foundation for cultivation. This method I propose for treating those soils, which consist of nearly pure vegetable matter. Of its success, I have no doubt. I regard this treatment as an exception to the usual rule, for I regard the practice of hauling clay to ameliorate a sandy soil, or the hauling of sand to ameliorate the clay, as generally too expensive. But in the case of peats and peaty soils, the vast quantity of fertilizing matter which they contain makes the improvement permanent, or at least quite lasting: the peat containing in itself lime, phosphate of lime, potash, soda, or those elements which are essential to all plants. The peaty soils furnish many kinds or varieties, depending upon the amount of vegetable matter they contain. The Open Grounds lie at the extreme, inasmuch as they consist entirely of this matter; especially near the surface. An addition of earths makes the most productive corn soils known. Some varieties of this soil I have examined. Of these, the two following are from Tyrrell county, taken from the new lands lying upon the Croatan sound: the adjacent ones being at least moderately productive.

One hundred grains gave 7.30 per cent. of inorganic matter. This 7.30 consisted of

Silex or sand,	6.02
Lime,	.02
Phosphate of lime, alumina and iron,	.90
Potash,	.20
Soda,	.06
Magnesia,	traces.
	7.20

The silex obtained, consisted of rounded grains, and was evidently sea sand. The per centage of organic matter is very large in this sample, but less than that of the Open Prairie. Still, this soil is productive for many years.

The cultivation of such soils, as the foregoing, results in the end, in the consumption of the vegetable matter; it is slowly oxydated, or, in other words, burnt. This chemical result, however, is essential to the growth of the plant. In process of time, so much of the vegetable matter is consumed, that the sand begins to appear among the vegetable matter; having become light, blows away and exposes more of the sand. At this stage, the productiveness diminishes. The land is found to be benefitted, but slightly, if any, by barn-yard manures, and it has become an important question, what shall be done, or how shall the fertility of such soils be kept up. On this question, there is but little doubt what that method should be: it is to add common soil from the road side, or from adjacent hills, and from the most accessible points. It is to be understood that they do not require organic matter; there is enough of that ; and inorganic matter furnishes the requisite quantity of phosphate of lime and potash, by its chemical changes. In this condition, is a part of a large field near Greeneville, in Pitt county, owned by Mr. Brown. It originally contained more inorganic matter than the soil from Tyrrell county ; but, in consequence of long cultivation, the sand begins to predominate, and the vegetable matter blows away. Decomposed slate rock would form the best application to such fields; next to which clay and common soil or marl mixed with earth.

Closely allied to the foregoing peaty soils are the two following: They were taken from the plantation of Captain Farrar, of Beaufort, Carteret county. No. 1, I found composed of the following substances :

Organic matter,	24.65
Water,	6.75
Silex,	55.37
Alumina,	7.62
Per oxyde of iron,	4.80.
Lime,	.40
Magnesia,	.29
Potash,	.09
Soda,	.03

The phosphoric acid was not determined. The soil was dried in the open air, and still it retained 6.75 per ct. of water, showing its retentiveness when charged with organic matter.

The other sample of soil, marked No. 4, was composed of

Organic matter,	24.94
Silex,	58.34

Alumina and per oxyde of iron,	10.49
Lime,	.27
Water,	4.47
Magnesia,	.15
Potash,	.06
Soda,	·05
	98.77

Both soils were taken from wet low grounds, recently reclaimed by ditching. They show the composition of good swamp lands, which, after reclamation by draining, will become the most valuable of all lands for corn.

The amount of organic matter varies constantly, and we find a gradual approach to the common soils, or those furnishing from three to six per cent. of organic matter. The two following are of an intermediate variety. They were taken from the plantation near Halifax. The first has been under culture for a long time.

Organic matter,	14.14
Water,	4.27
Silex,	78.64
Alumina,	1.69
Per oxyde of iron,	1.40
Lime,	.12
Magnesia,	.10
Potash,	.03
	100.39

The last is a dark brown soil, breaking into angular fragments or lumps after drying it. It is composed of

Water,	4.58
Organic matter,	12.00
Silex,	77.62

Alumina,	4.25
Lime,	.18
Magnesia,	trace
Potash,	.06
	98.69

Phosphoric acid undetermined.

Upon the Cape Fear River, I examined many localities where the soil would not differ from the two last. Among them should be ranked a large tract of land owned by Mr. E. M. McDowell. The soil alluded to has been partially drained, and put in part under culture. The organic matter varies from 12 to 24 per cent. It ranks among the best lands on the south side of the Cape Fear.

Where organic matter abounds, the soil is dark colored, and it requires no analysis to determine the fact. It will be observed by many, that there is no determination or separation of the coarse from the fine matters. The fact is, the soils consist entirely of fine matter ; it is frequently too fine : but fine materials are better than coasre, inasmuch as a coarse gravel, destitute of fine matter, is nearly barren, but a quantity of coarse with the fine improves the mechanical condition.

It may be interesting to give the analysis of two fertile soils, which cover a large portion of Southern Russia, and which stretch into Hungary. It is a vegetable mould, not much unlike some of those soils the composition of which has been given.*

Sand,	52.77	51.84
Silica,	18.65	17.80
Alumina,	8.85	8.90
Oxyde of iron,	5.33	5.47
Lime,	1.13	.87
Magnesia,	.067	.00

* Booth's Geological Report of Delaware, p. 127.

Water,	4.01	4.08	
Phosphoric acid,	0.46	.46	
Organic matter,	7.95	10.33	*Hermann.*

The insoluble matter in this analysis 71.42.

If this soil is taken as the standard for comparison with other soils, it should be observed that the quantity of alumina and iron might be diminished without impairing its fertility; and the quantity of insoluble matter also diminished or increased, without influencing its properties essentially; that is, it may contain 3 or 4 per cent. of silica, or less, and no perceptible change would follow from either. The quantity of lime is large, if compared with any soil in this country, unless we select a local one, resulting frem the disintegration of a limestone rock, or one composed of marl.

I may take this opportunity to remark, that we have no soils which can be called calcareous, or which are strictly entitled to that denomination. If we may appeal to observation and experiment, it is established that a small per centage of lime, only, is necessary to the highest degree of fertility; and yet this small per centage *is* necessary. If there is present one-half of one per cent. it seems to be sufficient; for it is rare to find a larger quantity in productive soils.

8. I have not yet examined the marshy lands of the Eastern Counties, adjacent to the sea, sounds, and rivers, as I design. I regard them, however, as among the most fertile and valuable lands of the State. It is upon these low bottoms, where the sluggish streams are partially dammed and obstructed by brush and logs, that the overflow of waters, bearing more or less of sediment, that accumulation of organic matter, takes place.

Centuries, however, have elapsed since this process began; and though the bottoms are composed of more sand than clay, still, they have accumulated from three to ten feet of vegetable matter; which, by the operation of chemical and physical causes, is being converted into the most productive of soils. In some instances, we have seen, as in the Open Prairies of Carteret, a growth of vegetable matter, instead of a deposit, as sediment; in others,—as in lands which are overflowed during the wet seasons of the year—a mixture of both; or growth commingled with sediments. The best lands are the latter—where a soil, with a magazine of food for plants, has accumulated, that nature has made the best provision for the planter. Notwithstanding a bountiful soil is thus made, still it should not be forgotten that it can be exhausted. The time, however, is so far ahead, that it seems almost needless to express our fears and caution to the proprietors of such lands. But yet the fertile lands of the West, which appeared to the early settlers as inexhaustible, are found now to have diminished in the burthen of their crops, and to require the application of fertilizers to bring them up to their original fertility.

9. The importance which I have given to the elements of soils, and the emphasis with which I have inculcated their several uses, may lead my readers to infer that it is sufficient to supply them, in order to secure large returns. I am, therefore, inclined to correct any misapprehension of this kind, by stating somewhat in detail, the mechanical or physical properties which soils should possess, in order to insure their fertility.

1. Fine and coarse materials.—A productive soil has usually a due admixture of coarse and fine materials. When it is all coarse, it is so porous that water passes through it too rapidly. A much larger quantity of fertilizing is required, and a large proportion is lost. Gravels have been called hungry because they consume so much manure: besides, there is little if any opportunity for chemical action. We must not lose sight of the chemical forces. Coarse materials cannot act upon each other.

2. The other extreme.—A very fine soil has defects almost equally great: when impalpable, it borders upon barrenness. The disadvantages of a fine soil are, water finds its way too slowly through it; it intercepts the free passage of air,

or, in other words, it is defective in its apparatus of circulation, and obstructs the growth of roots. If a seed is closely compacted in clay, it may be preserved for an indefinite time. For germination, air, moisture and heat are each necessary. In an impalpable soil, air is excluded. Manures applied to close soils or fine compact ones, are deprived of the air they require to fit them to act as food for the plant.

Taking the two extremes and comparing them together, we find the former consume manures, while the latter retain them unchanged: the first permit a rapid passage through them; the latter, the passage is obstructed and water is not allowed to enter. The impalpable and close soils are composed generally of clay; it is sometimes a pot clay. As a soil, it is dfficult to work: moreover, it is more difficult to secure the proper time for working it. In the Spring, it is too wet for a time, but it is slowly drying, and in process of time it will be in the best state for working, but it soon passes from a state of wet to one in which it is too dry. If worked when wet, it is injured for the whole season; indeed, it may be regarded as lost for the year, if it is plowed in a state when it is quite wet. But, if too dry, it is hard and lumpy. To cultivate such soils successfully, their constitutions must be well understood.

Ordinary clay soils may be managed by draining and the free use of lime. The forwardness in Spring of a wet clay soil, is a gain of two weeks by draining, and a diminution of one-fourth of the labor in tillage, and an increase of crop equal to one-half, and a total removal of the uncertainty in the time of plowing. The use of lime breaks it up, and imparts porosity and lets in air, moisture, and the free penetration of roots. The light colored clays are not so kind as the drab or yellow; the white are less productive. The common expression is, they are cold, and it may be cited as one instance, where a common expression is literally correct, for colored soils absorb more heat than white ones; their temperature is absolutely higher. The coloring matter, which is iron, diminishes the cohesion of clays; the colored ones are therefore less compact and far more productive.

From the foregoing remarks, it follows that the best soils are composed of both coarse and fine materials—coarse, that the air and moisture may penetrate them—fine, that chemical action and solution may be promoted, and the fertilizing matter retained. Oyster and clam shells are useful, even if unchanged, by promoting and preserving porosity in the midst of finely divided matter. The principles laid down are practical and easy of application, provided accurate observations are first made.

10. The mechanical properties of sand are directly opposite to those of clay. But I need not add to this head as it is too well known to require additional comment. There is a practice which should be recommended in cultivating sandy soil; it is to pass the heavy roller over them when moist. Seed sown upon the surface and then rolled, ensures its germination; if neglected, it is likely to fail. Light drifting sand may be fixed by the roller.* The comparative value of clayey and sandy soils have not been satisfactorily settled. The expense of working the former is always greater than that of the latter, but greater crops are usually obtained. Sandy soils have not stood so high as they deserve. Their easy tillage and middling crops, which may be obtained for a succession of years, pay well. They stand drouths better than clay soils, and admit of early tillage. A plantation is more valuable if it has both varieties of soil.

There are some facts, which go to show that argillaceous soils possess an affinity, if it may be so called, for ammonia, superior to a sandy soil. If so, it is one reason why argillaceous soils are better adapted to wheat than sandy, though not the only reason. The fact that clay absorbs ammonia, and requires a strong heat to set it free, is an important agricultural fact. In order to make the most of the absorbative power of argillaceous and other soils, fresh surfaces should be made by

NOTE.—The effect of the roller is confined to the surface. It does not press together the material below, which has been loosened by the plow. Soils then are not compacted by it; it does not destroy the effect of the plough.

ploughing, harrowing, or hoeing. New and fresh surfaces are much more efficient absorbers of ammonia than those which have remained unstirred for weeks. The late Professor Eaton was in the habit of illustrating this fact, and applying it to agriculture, by inserting fresh earth beneath a receiver filled with ammonia. The absorption took place almost instantly. The application of this fact to agriculture, was made as long ago as 1820. One of the advantages of ploughing and hoeing was attributed to the increased power of absorption of ammonia from the atmosphere, by the new surface thus made and exposed.

There seems to have been a special provision for furnishing a supply of ammonia, from which nitrogenous matters in grain are derived, and although ammonia forms an inconsiderable part of the atmosphere, yet bodies of various kinds absorb it with the greatest avidity, and in such quantites that it may be detected by chemical tests. I have already spoken of ammonia, (§ 3.) and have there suggested that the natural supply is adequate to the natural wants of the vegetable, and that for the purpose merely of preserving and sustaining all the species of plants now upon the earth, the atmospheric supply is sufficient. Even a luxuriant vegetation is sustained, and may be for an indefinite period, where herbage is mainly the product; yet, when the large products of grain are drawn from the soil, then the natural supply is insufficient, and the farmer is obliged to resort to artificial supplies.

MEANING OF THE WORD, IMPROVEMENT.—COMPOSITION OF SHELL MARL, &c.

§11. The soils of the eastern counties, it has been seen, furnish several distinct varieties, some of which lie at the extremes. The original constitution, which is sandy, aided by long cultivation, without due attention to the application of manures, has brough them to a condition, in many instances, of extreme poverty; and hence, it has become a question of great importance, how they shall be restored in a measure to their original fertility. This is not the only question, however, respecting the soils of the lower counties—how they shall be restored. Another comes up of equal if not greater importance, viz: How are the soils, which are now in a good condition, to be prevented from becoming poor and exhausted, and yet be subject to cultivation? Although we have presented two questions, yet, if either is answered, the other is also in the main; for the same principles are applicable to the two cases. The questions are not, how shall the crops be increased, for methods are at hand for doing this, without a permanent improvement of the soil. The crops of a plantation may be greatly increased by deep ploughing, and yet the soil is not virtually and essentially improved. Many are making a mistake in this respect. So, the system of clovering, or the use of green crops, might be followed out on a system combined with an alteration of crops. This, too, has been regarded as an improvement of the soil ; yet, it is not so, unless, indeed, it is accompanied with such additions of inorganic matter, which the soil require, and which are removed in the crops. The crops may be greatly increased without an improvement of the soil, and planters cannot learn this fact too soon. I do not object to the plan of increasing the crops for the seasons, by deep ploughing, subsoiling, and the use of green crops; but each and all, by themselves, cannot be regarded as an improvement of the soil There is something more, and it consists in the application, along with deep ploughing, subsoiling, and green crops, of all the elements which fertilize, and are necessary to supply the losses in the removed elements ; and it is only by pursuing this method that the spirit of the word *Improvement* will be realized.

In one sense, it is true, that any system, which adds a stock of essential elements to a soil, is an improvement; thus, by the use of green crops, or clover, or peas, we obtain from the atmosphere organic matter in the plants, which is ploughed in, and added to the soil. If the cultivation, however, goes on, this new accession of organic matter shortens the time during which the inorganic matter will last ; for more of the latter is used in the increased crops. There must be preserved a balance between the two kinds of matter : if, for example, there is too much organic matter in the straw of wheat, as there frequently is, when cultivated on new grounds, it is weak and falls down ; or else, there is an excessive development of the herbaceous part, and but little grain or seed.

But I shall leave this question to take care of itself for the present: it is time to speak of the real sources of improvement in the soils, as found in the lower counties:

The first substance requiring attention, is marl, a term which was originally applied to substances, which consisted in part of carbonate of lime; but, as oftener used, it includes calcareous clays, with or without shells. and argillaceous matters, containing silica, iron and potash, and probably, phosphate of lime, but destitute of carbonate of lime. The former are the marly clays, and shell marl—the calcareous matter is in the form of a carbonate ; the latter is the green sand, and contains potash, as its principal fertilizer, though it is now rendered highly probable that phosphate of lime is always present, and active in producing the results which follow from its use. In the green sand, however, there is no carbonate of lime, or but a trace, and hence, it may be better never to apply the term marl to the green sand, as it is so different in composition from the true marls, and so different in its geological position and age. But both are found in the part of the State of which I am speaking, one or the other being found in beds from Currituck to Brunswick, and from Wake to Carteret. The beds are not continued over very large areas: the green sands, however, are less isolated and more continuous than the shell marl beds.

12. The materials which are employed on the Atlantic slope, in Virginia, North and South Carolina, and other States still farther South, belong to two great sections or systems of rocks. The superior is the tertiary; the inferior system the cretaceous, occupying, in the latter, the lowest position in the system. It is that part known as the green sand, from the circumstance, that the beds are green or greenish, from the presence of numerous grains of silicate of iron and potash.

I propose to describe the first or tertiary beds. These, so far as my observation extends, are always isolated, or confined comparatively within narrow limits. They are not spread out so as to form a continuous bed; but limited usually to a few acres, perhaps many acres, and completely disconnected or separated from other beds. This view of them is important, inasmuch as it does not follow, that, because a bed appears in a branch, on one side of the plantation, that it will be found on the other side of it, though very desirable that it should. Some beds are confined to an area of an acre. Some are but a few rods square, and others are still smaller, and appear like nests of shells in the midst of sands. The beds of oysters and clams are, indeed, good representatives of marl beds, as to extent: some larger, others smaller. If we examine the bed or floor of the ocean, by soundings, we shall find it composed of materials, especially along the coast, very much the same; but its surface is not evenly spread out. In some places it is smooth and level; in others, it rises in ridges and hills, with their vallies. This disposition of the materials, forming the ocean's bottom, provides, if it may be so called, a variety of climates—some adapted to the wants of living beings; others incompatible with life. Some are sheltered, and others are exposed to the lashing of the waves. It is in these sheltered abodes that we find life in its various conditions and stages of development. While upon shores and in soundings, where the waves and the elements are at strife, life is absent, from its exposures. So, when the beings whose remains constitute and form these marl beds, peopled the waters, there were sheltered places, quiet and still bays, which favored the development of life, and it is upon such areas that these deposits were made, while other areas, exposed to sudden changes, separated those, teeming with life, from each other. We have reason to infer, then, from observations upon the ocean's bottom, that the areas of the marl beds would not be found spread out continuously; though marl beds, possessing characters in common, furnishing the same kinds of shells, will occur at wide and distant points. Not only, too, are the beds characterized by similarity of forms and kinds, but the accompanying sediments, sediments of the same mineral character, would be found with them. This would be necessary: it is one of the provisions of life—the medium which conveys their food and the habits and habitants must and should agree.

13. We reason, then, from life to things, and things to life. Wherever the conditions for the life of the clam and oyster were favorable, or, to be more general, where the canditions of life were favorable to a larger number of Molusca, there they would be congregated, because this food, the climate and all, would conspire to favor development and growth. Similarity of organic forms, then, become indicative of the value of marl deposits, over wide and extended areas. Marls which contain similar shells will be found to possess nearly the same agricultural value.

14. The marls are distinguished by different names in the vicinity where they occur. The red, blue, and shell marl are names applied to beds occupying the same geological positions. Sometimes there are some differences in their properties and value. The red marl owes its color to a change in the oxyde of iron mingled with the shells. It has changed from a state of protoxyde to the peroxyde. It is due to exposure to the atmosphere, and is usually the superior part of the bed which has under-

gone this change. The blue marl still holds the iron in a state of protoxyde, which imparts a bluish green color to the mass. The term blue marl, however, is frequently given to the green sand, an inferior and older formation, and which owes its fertilizing properties to potash, as I have already had occasion to say. We might make a distinction between the sandy marls and the argillaceous. In the first, sand predominates: in the other, a bluish clay. Both effervesce with acids;—the latter is the most valuable. The proportion of carbonate of lime is variable; or, what would amount to the same thing, the amount of sand is variable in the same bed, and in the distant beds which occupy the same position; though to the lime is due the existence of the animal which inhabited the shells.

§ 15. The marls of Cape Fear river furnish all the varieties which have been noticed in the foregoing paragraphs. The first beds which appear, upon the river, are about ten miles above Elizabeth, in Bladen County.

Mr. Lassaine's, which is the highest point visited from Elizabeth, is sandy; Mr, Gillespie's is argillaceous; and Mr. Cromarty's is more calcareous, and parts of it are cemented together. It is a mass of shells, and has been found, by experience, a valuable fertilizer. It is seven feet thick, and underlies many acres.

Mr. Cromarty's marl yields

Silex,	52.50
Alumina, phosphate of lime and iron,	7.15
Carbonate of lime,	40.50
Magnesia,	.75
Potash and soda	trace
	100.90

§ 16. The quantity of marl containing the per centage of lime given above, requires, per acre, for soils not remarkably sandy, about 200 bushels. The experience of planters is, that very poor soils are injured for a year or more by the application of marl, except in small quantities. One hundred bushels is regarded as sufficient for sandy exhausted lands. When two or three hundred has been used per acre, the land is said to be burnt, or the vegetation is, in part, destroyed; and the practice is to begin with the lowest quantity, and proceed in marling by subsequent additions; being governed by the quantity of organic matter restored to the soil. Many planters have observed that heavy marling is injurious to poor lands, who do not attempt to give a reason for the statement. If the common opinion respecting the danger of applying too much marl to poor soils is founded on correct principles; or if there are lands upon which it would be hazardous to apply it in large quantities at first, we may be assured that it will be safe, always, provided it is mixed with much organic matter. The prior mixture and incorporation of the materials with leaves, bark, decayed wood, rich loam, peat, &c. obviates the objection raised. The practice in New Jersey is regarded as the best:—namely, the prior mixing of marl and vegetables. It is true that the Jersey marl is destitute of lime. Probably the great danger of bringing the use of marl into disrepute, by representing its injurious effects upon poor soils, has more frequently arisen from too high expectation of receiving great effects the first season that it is applied:—whereas, the better and safer course is to bring the land back gradually to a good standard of fertility; pursuing that course which is calculated to increase the vegetable matter in the soil for several successive seasons. A plan like the following is deserving of trial: Spread upon an acre seventy-five bushels of the fifty per cent. marl, and put it down in peas. When in blossom, plough in the crop, and sow rye or millet for the succeeding crop. The land will have gained a sufficient amount of organic or vegetable matter to admit of the use of one hundred and fifty or two hundred bushels at the next marling time. Some land will require the loss of two crops, perhaps, before they can be treated with a fresh dose of marl. The doctrine to be inculcated is, to exercise patience with light and worn out soils, and not expect too much at first, when the first step towards their fertilization is taken. When the work has been properly conducted, the planter may regard such lands as so much added to his possessions, of durable and productive fields.

Abundance of the shelly marl lies in the bank about one half a mile, probably less, below Elizabeth. It forms a stratum from two to three feet thick, in the bank upon the south side of the river. Coprolites and teeth of fish are common. The latter are mixed in the bed with the shells, more or less. Both teeth and coprolites lie at the bottom of the structure, intermixed with some bones, and rounded pebbles of quartz. This layer at the bottom, intermixed with pebbles and rolled coprolites, is an interesting feature of the bed. I have been in hopes that in this position, in some favored place, coprolites, in sufficient quantity, might be discovered, to pay the expense of extracting them separately. They possess a composition superior to bones, and may be used for the same purposes as bones.

The following results of an analysis represent, in the main, their composition :—

Silica,	9.68
Phosphate of lime,	71.59
Carbonate of lime,	11.28
Magnesia,	.50
Potash,	a trace
Organic matter and water,	4.40
	97.35

The coprolites of this bed are all black, or dark brown. They are quite hard, and may easily be mistaken for the dark pebbles of quartz, with which they are associated. They are generally broken, and are rounded; but some retain their original spiral form. They are two and a half to three inches long, and three-fourths of an inch in diameter.

§ 17 Below Elizabeth in Bladen county, the marls continue to be exposed at intervals. One of these exposures is Walker's bluff, nine miles below Elizabeth. It is the highest upon the river. It presents a steep escarpment, which consists of different colored sands, with a thick layer of shelly marl. The marl is also more or less sandy. Eighteen miles below Elizabeth, the bluffs appear upon the river, with their strata of sands and marls. The strata are also well exposed at Mr. Robinson's plantation, one mile above Mr. Brown's landing. The following strata appear in the banks at Robinson's, beginning at the top: 1, twenty feet of different colored sands, some yellow, brown and white; 2, twenty feet of blue marl, more or less sandy, and calcareous at the bottom; 3, a single layer of blue compact clay, 8 inches; 4, sand; 5, yellow and brown sands: 6, blue marl, containing a single species of ostrea. Most of this stratum is below water, and hence its thickness is not determinable by inspection. The marl bed is very thick, but contains considerable sand in its superior part; yet it is found a valuable fertilizer.

The marl stratum, at Brown's landing, is three feet thick, and contains many shells and much green sand, in grains, and seems to have derived its materials from the green sand of the cretaceous formation below. At Mr. McDowell's, the green or blue marl appears in a low bank, one mile from the river. Also, on the plantation of Messrs Andrews. These beds are peculiar in their geological relations, and merit a careful examination.

Ten miles below Mr. Brown's landing, at Black Rock, the shell marl appears in the bank, but is quite sandy, and appears as if this stratum is discontinued, and ceases at or near this place. It is scarcely more than one foot in thickness. Immediately below it, the green sand is well developed, and it is well characterized by its fossils.

On the road from Brown's landing, to Black Rock, beds of marl appear, which are evidently isolated. The facts all go to show that the strata of shell marl never form very extensive beds: even that so conspicuous at Walker's bluff, disappears suddenly, and its place is taken by the different colored sands.

§18. The strata of marl, which I have thus far spoken of, are composed of many kinds of materials, intermixed irregularly with each other. They possess many fossils in common, but often rare kinds are found, in one or more of the beds, which is not generally distributed. But again, there are many places where the oyster shell is the principal one, and which, instead of crumbling in the hand, and by its own weight, are firm and nearly as sound as those now living upon their beds. The value of oysters, in this condition, is far less for immediate use, than those which are decomposed: indeed, for spreading upon the soil, the principal effect must be mechanical. If, however, five hundred bushels were used per acre, good effects might be expected; for there is a slow disintegration, and there is a slow solvent action, also, by which lime will be given to the soil. Of this character, are those shell banks immediately upon the coast. These, though they have been exposed to atmospheric agencies for a much less time than those in the interior, are nevertheless, farther advanced in the process of decay The best method of employing the undecomposed shells, will be to burn them; use the quick lime, or after it has passed into a sub-caustic state.

§ 19 The Neuse valley is deeper and lower than the Cape Fear, and hence it furnishes a larger supply of Marl beds. The Chapony Hills have been known for a quarter of a century, to be rich in marls of different kinds The vicinity of Goldsboro,' however, possesses most distinctly the characters of those upon Cape Fear. The beds which are best known are upon the plantation of Messrs. Scott, Ham and Peacock. The beds are identical in age and position, and belong the middle tertiary; they are from twelve to fifteen feet thick. These shells are embedded in a green marly clay, which effervesces with acids. Mr.

Ham's marl is filled with small shells, which have so far decayed that it is difficult to find one entire. The covering to the different beds is quite varied. Mr. Ham's has three feet of peat, which is probably the best substance, considered economically, which could have been placed there; it is the very material wanted to secure the best effects of the marl, and to form with it a compost. It is not determined what strata lies below these beds, occupying, as they do, grounds which are low and depressed. The marl of Mr. Ham's may be regarded as composed of—

Sand or Silex,	45.60
Phosphate of lime, per oxyde of iron and alumina,	8.25
Carbonate of lime,	44.15
Water and organic matter,	1.60
	99.60

The marls, previous to analysis, have become dry by exposure to the air. Some moisture and organic matter remains, varying from one to three and four per cent. The sand is always greater than appears from simple inspection, and it usually consists of fine grains of pure quartz. There is also, one-half of one per cent. of soluble silica, which is usually omitted.

§ 19. It will be observed, that in making up a statement of the analysis, I place the ammoniacal precipitate, the oxyde of iron and alumina, under the head of phosphate of lime, instead of placing it in analysis under the head of alumina and per oxyde of iron. I have done this, because this precipitate consists mainly of phosphates, though the exact amount of phosphoric acid has not been fixed with accuracy; yet, one-fourth of a grain of it gives a strong rendition of phosphoric acid with molybdate of ammonia.

On the banks of the Sarpony hills, on Mr. Griswold's plantation, marl of an excellent quality, and in great abundance, exists. The beds, however, are indurated, or have passed into that condition which is known as stone marl.

The following is a correct description of a section of a slope or bank, where excavations have been made for procuring limestone; beginning at the water's edge:

1. Stratum of marl extending beneath the water of the Neuse, in a soft condition.
2. Consolidated marl.
3. Sandy marl.
4. Granular and partially indurated marl.
5. Stone marl fifteen feet thick, and which has been used for lime.
6. Sand.

The whole bank has a thickness of thirty or thirty-five feet. It is one of the best locations on the river for the manufacture of lime for agricultural purposes, and it is not a little remarkable, that property, which might have been very valuable, and at the same time useful, to a whole community, has been lying useless and unproductive.

§ 20. At the Sarpony Bluff, the formation presents an interesting section to the Geologist. It would be expected that the marl would appear here, as at Walker's Bluff, on the Cape Fear; inasmuch as the height and formations do not materially differ. The Sarpony Bluff is between 75 and 80 feet high, and consists of the following strata:

1. Sand extending beneath the water, 4 feet.
2. Band of pebbles and sand, cemented by iron, with casts of obscure vegetable stems, five feet.
3. Gray sand, thirty feet.
4. Ferruginous band, eight feet,
5. Light colored ferruginous layer.
6. Copperas beds, consisting of pyrites, clay and vegetable matter, nearly black. It is properly a bed of lignite, charged with pyrites.

7. Sand, twenty-five feet.
8. Earth, sand, &c. compacted together.

These beds, it will be observed, are mostly ferruginous, or those which are highly charged with the oxyde of iron; and it should be observed, that, where iron is thus in excess, the beds do not furnish animal remains, or marl beds. Fossils are rarely distributed in them;—sulphuret of iron is usually the source of the oxyde, in beds of this description, and, in decomposing, forms an astringent salt of protosulphate of iron or copperas. The marl of Griswold's plantation thins out before it reaches this high bluff; a change which occurs also on the Cape Fear, where the marl suddenly disappears, being replaced by sand.

§ 21. The vicinity of Newbern has long been known as abounding in marl. New beds are frequently brought to light by accident, and sometimes by careful exploration of favorable places. Judge Donnell, during the past year, has discovered shelly marl upon an old plantation;—proving that most plantations, which are elevated considerably above the river, are not destitute of this fertilizer.

§ 22. The Tau river, in its banks and branches, is rich in marls of the age of the middle tertiary, adopting the views of the Geologists who have examined, with some care, the fossils peculiar to these beds.

§ 23. Beginning in Nash County, five or six miles above Rocky Mount, we find the shelly marl at intervals as far down as Washington.

The first I shall notice is from Mr. McDaniel's, 5 or 6 miles above Rocky Mount. This marl, like many other kinds whose quality is equal to the average, is more or less consolidated, and breaks up into masses. Thin laminœ of coal, or lignite, are mixed with the shells—a fact which indicates that the source of the earthy material was in the coal formation, in part. This marl is regarded as consisting of the two kinds:—the brown or red, and the blue. Practically, I think it well to keep up this distinction; for the red, thus far, has given better results in analysis than the blue. I do not know what opinions are entertained by planters of their comparative value, who use both kinds.

The analysis of two specimens of this marl gives very good results for the red variety:

Silex, or sand,	16.25
Phosphate of lime and per oxyde of iron,	10.00
Carbonate of lime,	71.75
Organic matter and water,	2.15
	100.15

The magnesia and potash were not sought for.

The appearance of this marl is quite unpromising, as it is quite bumpy and hard, passing into an indurated marl. Still, analysis shows it to be an excellent kind, and which, I am confident, would yield 7 or 8 per cent. of phosphates, over and above the alumina.

The blue marl which is found below, gives a good analysis, but contains less lime:

Sand, or silex,	21.25
Phosphate of lime, and per oxyde of iron and alumina,	10.00
Carbonate of lime,	64.65
Organic matter and water,	2.10
	98.00

These marls, when tested, have always furnished a small quantity of magnesia, and a trace, and sometimes a weighable quantity, of potash. The two samples furnish the same amount of the phosphates, and oxyde of iron. The color of the ammoniacal precipitate is darker in the red, than in the green variety, indicating a larger quantity of the oxyde of iron.

This bed, which furnished the foregoing samples of marl for analysis, is the highest known to me upon the Tau River. This fact, however, does not prove its non-existence still farther; and I predict that careful examination will reward the planters in that county, with many additional beds. Every bed must be regarded as a prize, if it is limited to 50 loads.

§ 24. The reputation of marl, as a fertilizer, in Edgecombe County, has led most of the planters to search for i; upon their premises. Probably there is no better proof of the value of this substance than is furnished by the estimation in which it is held by its citizens. Regarded, in years which are past, as a county somewhat behind the times in literature and science, she has, nevertheless, outstripped all other counties in the application of good sense and common sense to her farming interests. Facts are sometimes misunderstood, as well as misrepresented abroad, when applied to the internal policy of a State. So, I suppose, Edgecombe has been misunderstood;—for agricultural improvements are incompatible with ignorance and darkness. If we find a people alive to their internal interests, so vital as agriculture, we may be sure that mind has been at work. But, however this may be, Edgecombe has the reputation of being the first county in its agricultural improvements and agricultural prosperity. Her success has been secured chiefly by her marl beds; it would have been secured, in the end, if marl had not existed; but more time and capital would have been required to have placed her in her present enviable position.

Although the foregoing remarks may be regarded as out of place and uncalled for, yet I deemed it right to give credit where it was so justly due; without at all questioning the ability of her neighbors to compete successfully with her for the next five years.

There is another fact worth recording:—Edgecombe has many men who have been educated at her excellent University, who regard agriculture a befitting profession for an educated man—an example which the friends of agriculture will be pleased to see imitated in other parts of this Republic.

§ 25. The Marl beds at Rocky Mount belong to the same age as the preceding. They are the blue shelly beds frequently furnishing that large scollop or feature, which is regarded as characteristic of the middle tertiary. The appearance of granite and sienite at Rocky Mount, has produced a series of falls in the Tau river; and sometimes the marl is found resting immediately upon those pyro-crystalline rocks. The beds are associated with the following strata:

1. Above the marl, stratum of sand and rounded pebbles, which is ten feet thick.
2. Marl somewhat sandy, but impervious to water, and hence, the surface water percolates through the upper mass and is thrown out by the marl. The upper is made up of fine or small shells, like that of Mr. Ham's of Goldsboro'. The lower is intermixed with the large scollops and clams—(*Venus difformis.*)

The marl, like that of other beds, is rich in lime, and often consolidated or cemented in different parts of the structure. The whole thickness of the shelly strata is seven feet. The marl is sometimes charged with rounded pebbles of different sizes. The position of the marl is upon the banks of the Tau; several beds appearing in the banks near the falls, or at one-half, and also, about one mile, below the railroad bridge. There are points where excavations have been made, but it is probably continuous for nearly a mile. Whenever there is an undulation by which the strata are elevated even a few feet, there the marl appears in the banks. Rounded stone and pebbles are strewed over the surface in great abundance, but this fact is no indication that currents have swept over the country in a certain direction.· Some of the soil at Rocky Mount is light and requires the application of marl to give it more retentiveness, as well as to furnish a fertilizer to supply the waste to which the lands have been subjected.

The marl strata reappear at Tarboro', at many points; sometimes on the river banks, and sometimes in the banks of creeks. One of the important beds is near the village, and belongs to Mr. Bullock.

The section which contains the marl, is made up of—

1. Sand which extends below the water of the creek.
2. Clay with lignite, three or four feet.
3. Marl, seven or eight feet.
4. Sand and clay without fossil, or only a few casts.
5. Sand, gravel and soil.

The marl is intermixed with coprolites, a few bones, and water-worn pebbles—mostly at the bottom of the bed. There is the same tendency to consolidation as at Rocky Mount, and at other places on the Neuse and Cape Fear rivers: The same shells, consisting of large pectens, *(Pecten Madisonius,) Venus Difformis*, and two or three species of Pectunculus. Masses of sulphuret of iron are not uncommon.

The marl of this bed is composed of—

Sand or silex,	56.25
Phosphate of lime and oxide of iron and alumina,	7.50
Carbonate of iron,	34.15
Organic matter and water,	2.10
Magnesia,	.56
	100 56

Mr. Bridge' Marl.

It will be observed, that rather more than one-half must be set down as useless matter. The analysis was made of that portion containing the small bivalve shells, and as many of the shells are rejected as convenient; there will, therefore, be more lime than is given in the analysis, by three or four per cent. It is, perhaps, unnecessary, to remark, that the finer the material the better; that the marl with small bivalves, is better than the marl with large ones. The latter when abundant is better for quick lime.

Mr. Knight's marl bed is three miles from the village, and has been extensively employed in marling: It is upon the banks of the Tau.

I obtained the following section of its beds:

1. Sand and gravel at the river's edge.
2. Sandy marl.
3. Marl with shell, six feet.
4. Greenish or blue clay, six feet, containing casts of shells only.
5. Sand.

The whole thickness is about thirty feet.

This bed has furnished many large bones, both of Saurians and land quadruped, principally of the Mastodon. This bed has been regarded as equal to the best of the varieties of shell marl. Sand seems to be a constant associate of the marls. It occurs both above and below the stratum of shells. In this respect there is a general uniformity in the marl deposits in the different vallies—the Cape Fear, tht Neuse, and the Tau. The intermixture of sand is the material which diminishes or changes this value. Though coarse shells, as the large scollops and clams, together with certain species of oyster, constitute a poor kind of marl—these resist for a long time the action of the weather.—Where these have abounded, I have heard unfavorable reports of the effects upon the soil;—or, at least, the good and advantage expected were not realized. This all goes to show the importance of a comminution of the material:

it favors solubility. Those agents, as water and carbonic acid, act with more energy, and the power of absorption is increased in the substances themselves.

§ 27. Where the coarser marls are necessarily employed, the advantages of a crusher is obvious. Plaster is operative immediately, because it is ground fine ; if it were more in the condition of coarse shot, its effects would not be apparent on most of soils. The subject of comminution is one of considerable interest in husbandry. It is not expected, however, that soils can be ground or comminuted, except through and by the action of the weather. The marls which are coarse, however, when made into composts, will be improved materially, especially when these composts are composed of organic matter, which liberates carbonic acid. Frequent stirring is also important. Another mode is by the application of marl. Exfoliation of the large shells begins at once ; the loss of organic matter is replaced by water, and the whole becomes porous. One fact worthy of notice, is, that mixtures are always more valuable than simple bodies ; even phosphate of lime is more active and beneficial when intermixed with materials constituting a compost, or intermingled with a compost. The constitution of man and animals requires mixture. We have seen that the soil is eminently a compound mass ; and when food is taken into the stomach, there are agents which assist its reception in large quantities into the system. So long as we have regard to the necessities of plants, we can hardly form a mass of compost, too complex in its constitution, or which shall consist of too many elements, and I think it highly probable that many failures have arisen from neglecting the aid to be derived from intermixture.

§ 26. The marl of Mr. Bullock's, near Tarboro', and upon his home plantation, has been fairly tested, and proves valuable.

The section of the slope in which it occurs, is represented by the following beds, beginning with the lowest :

1. Sand.
2. Marl, with shells, scollops, &c., $3\frac{1}{2}$ feet.
3. Blue compact clay, which contains decomposing pyrites.
4. Sand and clay, in alternating layers, mostly destitute of fossils, 5 feet.
5. Sand.

The blue or greenish marl of Mr. Bullock's plantation has the following composition:

Sand,	34.40	
Phosphate of lime and oxide of iron,	3.20	
Carbonate of lime,	54.52	
Magnesia,	1.50	
Potash,		trace.
Soda,		trace.
Organic matter,	4.88	
Water,	1.38	
	99.88	

Mr. Bullock's plantation consists of rather more than one thousand acres. It lies in a great bend of the Tau river. From the river, to the higher ground, there are four distinct but low terraces. The average crop of seed cotton is about twelve hundred pounds. The marl is, in part, composted; it is, however, allowed to be exposed to the weather, and undergoes certain mechanical as well as chemical changes, prior to use. Probably, it is always important to give the marl air, as it may be termed, before it is spread upon the soil, even if no mechanical change is effected by it.

Marl, which is a year old, is much better than when taken from the pit, and spread immediately upon the soil, especially if it is turned over three or four times during the year.

§ 28 The improvements of the Panola plantation, under the direct supervision of its intelligent proprietors, Messrs. Norfleet & Dancy, exhibit something of the spirit which pervades Edgecombe.

The plantation was old, and was purchased for $65 per acre, and consists of 908 acres, 550 of which is now under cultivation. Its former proprietor had pursued the system of rest so common in the South, without a thought of providing for the future, when the most valuable parts of the soil had been converted into corn, cotton and bacon, and sold in a distant market. Its new proprietors, on making this purchase, were aware that the old system could not be pursued, and they were well satisfied that the only system which could renovate the soil, though originally good, was to supply an abundance of fertilizers or manures. The plantation rises in three or four terraces from the river, the lowest of which is often overflown with the high water of the river. Logs, flood wood and trash cover the lower terrace, and occupy the low ravines. By a judicious application of the force of only two laborers, three thousand bushels of ashes were made in two weeks from this refuse wood. In addition to this important fertilizer, twenty thousand loads of compost were made, consisting of cotton seed, stable manure and river sediment, and the muck of ditches. Ample manures were taken for draining, by a free opening and deepening of the old ditches. The main body of the land is rolling, the higher parts are sandy, and the lower formed of a clay loam.

The points worthy of notice, are the preparations for a productive farming, and the expenditure of capital for this purpose; and, although it would seem, that the plantation itself had furnished a large amount of material, at a trifling cost, still, bones and guano were also prepared at a cost of $52 per ton, and bone dust, at fifty cents per bushel, in New York.

§ 29. The first and important lesson, which the agriculturist should learn, is, that he must supply his land with manure, and if any planter will calculate the cost of a full supply of manure, and then the cost of new clearings, required by the old system of husbandry, he will find it cheaper, and hence, more economical to make and buy manures, than to clear up his plantation, for the purpose of cultivating new lands, and those which have been partially restored by rest. The improvements of the Panola plantation do not terminate in furnishing an ample supply of manures. The removal of the cabins to an airy, healthy and central position, is one of the most important improvements. The arrangements, too, of the out-houses and water sinks, so as to save nitrogenous matter, with their phosphates, is another step in improvement, worthy of imitation by others. So, also, it is made the special business of some one or two laborers, to collect all matters which may be used as a fertilizer. But I need not dwell upon other minutiæ of the improvements designed to secure, in the end, a profitable investment of capital. Considered in the light of a speculation only, it does not require a prophet's vision to predict the result.

In the foregoing remarks, I have had in view the fact, that information of what others are doing is one of the best stimulants to improvement by others. The most important results will be brought about by the successful projects of enterprising men, when they are made known. It is a principle which applies to all professions.

Now, the season having passed, and the crops been gathered and weighed, it turns out that the cotton fields have yielded one bale of cotton, of four hundred pounds, to the acre, which the year before did not amount to one half of that, and the corn lands, which, before the improvement, would not and did not yield three barrels to the acre, have yielded, this year, eight: a well marked and decided improvement. The season, it is true, has been favorable, and it should be noticed in making up the results.

§ 30 I have one more remark to make in this connexion: it relates to the effect on the product, when high cultivation is resorted to. This effect is of the highest consequence, and it does not end with a simple increase of product, but also in a product of a better quality. We probably, however, understand the mode of increasing a productio n, better than giving it a superior quality. The lint of cotton is better, if produced by high cultivation, than by an indifferent cultivation. Indian corn is better, when the land is supplied sufficiently with its proper food. It is light, if it lacks food in the soil. Wheat is heavier, by three or four pounds to the bushel, if grown on a rich soil. Barley is sold by weight, for different soils produce a grain lighter and more chaffy than others. Oats vary much in their weight, by being grown on soils differing in their fertility.

New lands are productive, and at the same time give a superior quality of grain. On old lands, there is a diminution of weight, and a loss in the quality of the product: there is more offal. Attention should be given, then, to the quality of the cotton, as well as to the quantity. The planter may control, in a manner, both results, or, in other words, he may modify results, by cultivation. It is well known that cotton requires a stiffer soil than corn. The principles involved in a cultivation of these two staples of the South, are not the same. The object, in the cultivation of Indian corn, is the development of cellular tissue. I do not yet know the precise modes by which we can apply principles successfully to practice. Yet, the cellular tissue requires, for its development, more carbonate of lime than phosphate of lime. Analysis of the different tissues proves this. If this is true, it is an indication that the marls are adapted, especially, to the growth of cotton; that while it contains some phosphate of lime, as this is necessary to all tissues, yet the lime in the cellular tissue is furnished, originally, from the carbonate.

Experiments might be devised for testing the truth of these views;—the object being to increase the lint, and improve its quality. Has any attention been given to the selection of seed?—selecting from the field the seed which has first ripened, and which has given the longest, finest, and most silky staple?

The marl beds of the Tau River are exposed at points below Tarboro', from Greenville to Washington.

§ 31. At Greenville they have been successfully used—it belongs to the middle tertiary. Just below Sparta, the left bank is thirty feet high, and there is exposed a remarkable stratum of marl. Above Sparta, the bank is too low to expose it.

In the vicinity of Greenville, the marl beds are numerous. Mr. Brown's bed exhibits the following strata:—

1. Sand exposed at the bottom.
2. Two feet of sandy clay.
3. Three inches of yellow sand.
4. Eight feet of shell marl, with greenish grains.
5. Sand, with sandy clay, of a green color.

Mr. Britton's marl exhibits a section quite similar to the above:—

1. Green indurated sand.
2. Marl, six to seven feet thick.
3. Sandy Marl, one foot.
4. Brick clay, four or five feet thick.
5. Sand.

This marl is reddish, and operates favorably and quickly. The stratum of clay occupying this position is not uncommon. In fact, it is almost continuous over the whole country, though it is not always present as a covering to the marl.

A bed on the plantation of Mr. Boyd, in the same neighborhood, is about fifteen feet thick: it is overlaid by a band of yellow clay, upon which there is sand five feet thick.

§ 32. Six miles below Greenville is Dr. Dixon's marl bed, which had just been opened at the time of my visit. It is blue shelly marl; most of the shells are small; and the mass is much disintegrated.

The strata lie in the following order:—

1. Marl 15 feet thick—its bottom not certainly exposed.
2. Blue clay, 3 inches.
3. White loose sand, differing but little from drifting sand.

This marl is composed of the following proportions in fifty grains:—

Sand,	15.70
Carbonate of lime,	27.30
Phosphate of lime and oxyde of iron,	1.60
Water,	1.69
Magnesia,	.11
Potash,	trace.
Organic matter,	2.94
	49.34

In the banks of the Tau, at Greenville, numerous flattened masses are washed out of the bank. The color is a drab, or light yellowish brown. They are frequently perforated by a round hole; they have a close resemblance to the ordinary clay stones. Coprolites are associated with them: and I was inclined to regard them all as coprolites:

but it proved that many of the flattened bodies are not coprolites. Analysis of one of them gave the following results :—

Insoluble matter,	.13
Phosphate of lime,	14.50
Carbonate of lime,	10.50
Magnesia,	trace.
	24.13

The coprolites have always given potash, when tests are applied. These substances in the Greenville beds are soft, and unlike coprolites which occur on the Cape Fear river. They are unlike them in color and form. Most of them are, in their flattened cakes, not much unlike a cracker in form; though, in this respect, there is much diversity.

The country around Washington is too low to give good exposures of shell marl. It is, however, common in the low banks, but liable to be overflowed.

§ 33. Mr. Myers' marl bed gives the following section:

1. Blue marl.
2. Shelly marl, 3 feet.
3. Red marl, 8 inches.
4. Brick clay,
5. Sand.

Another bed, upon the plantation of the Sheriff of the County, was too much concealed by water at the time of my visit. A specimen of the marl furnished for analysis gave the following proportions:

Water,	1.40
Organic matter,	2.70

Sand,	28.30
Phosphate of lime and oxyde of iron,	5.13
Lime,	10.81
Magnesia,	.11
	48.45

The analysis contains less lime than was expected. The shelly portions were rejected in part; which, had they been included, would have given a larger per centage of lime. The effects, as they have appeared upon trial, were remarkably good and satisfactory. The absence of high banks increases the labor and expense of raising the marl.

I took occasion to visit Jones County, on my return from the examination of the State lands in Carteret. The Hon. Mr. Donnell, of Newbern, accompanied me, and laid me under many obligations for the information received of the country.

This County has an undulating surface; the soil has more clay than Edgecombe or Pitt. The foundation for the highest improvement in agriculture exists in its soil. Less cotton is cultivated than in Edgecombe; but, when cultivated, it is not difficult to raise it up to sixteen hundred pounds of seed cotton per acre. Marl of a peculiar kind exists in the waters of Rainbow Creek, and on the banks of Miller's Creek. The marl is formed of the debris of exceeding large oyster shells, some of which are 14 inches long, and 1½ inches thick. They sometimes weigh 6 and 7 pounds. The surface shells are decomposing; those deep in the beds are quite sound. The marl, however, of these beds, is less valuable than when composed of small shells. The testimony of those who have been acquainted with its use is of a negative kind; but still I could not learn all the circumstances attending its application. At Pollocksville, on the Trent, this marl appears in its banks, and presents the following section:—

1. Sand.
2. Oyster bed.
3. Sand.
4. Oyster bed.
5. Sand.

It is about 20 feet to the second bed of oysters. Beneath these beds is the lime rock of the country, consisting of consolidated marl, having the same characters as that upon the Trent near Newbern. In many places, its purity is such that it makes a good lime; in others it is sandy, and makes a weak lime.

§ 34. The marl of Little Contentney Creek possesses the same characteristics as that of the Tau and Neuse.

For the opportunity for making the examination of Little Contentney, Tossnot, and a part of Nash County, I am indebted to the kindness of Mr. Myers, of Washington, President of the Greenville and Raleigh Plank Road.

The marl upon the plantation of Mr. Streeter was too much concealed by water to admit only a slight examination. The fossils, however, proved the deposits to be of the middle tertiary. The large Pectunculus and Venus difformis, common at other places, were observed among other common fossils of the formation.

The beds upon the plantation of Mr. May were also covered with water. These, in part, were sandy, and a specimen gave only a small per centage of lime in the analysis.

As for example:—

Sand and silica,	81·20
Phosphate of lime and oxyde of iron,	8·00
Magnesia,	trace.
Carbonate of lime,	5.60
Water,	1.20
Organic matter,	2.60
Potash,	trace.
	98.60

This marl, as poor as it is, containing less than twenty per cent. of available matter, has increased the crops, according to the statement of Mr. May, fourfold. It is probable, however, that this sample is not an average of the marl stratum. The soil of Mr. May is sandy—at least on parts of the plantation.

§ 35. The marl of Col. Barnes, upon the Tossnot, is similar to that upon the plantation of Mr. Ham, near Goldsboro'. It is the blue marl, intermixed with innumerable small bivalve shells, which have become very thoroughly decomposed. The bed is eleven feet thick, covered with a stratum of sand five feet thick.

§ 36. The deposits of marl upon the Roanoke are no less important than upon the Tau, Neuse and Cape Fear. My examinations were confined chiefly to Halifax County. The beds, considered as one formation, consist of the following members :—

1. Layers of decomposed rock—a coarse mica slate.
2. Marl loaded with fossils, five feet.
3. Marl of a green color, with only a few shells, eight to ten feet.
4. Blue clay, from ten to fifteen feet thick.
5. Reddish clay, two feet.
6. Gravel, fine and coarse, twenty feet.
7. Gray sands and loam.

The marl lies deep, and is exposed only in ravines. It is attended with much expense in raising it. Mr. Pope, of Halifax, has used it upon his plantation, and has made preparations for its extensive consumption, and the results have been favorable. The soils of Halifax, having been under cultivation a century and a half, or more than a century, have become exhausted.

The soil of one of the oldest plantations gave the following results on analysis :—

Silex or sand,	95.38
Alumina and oxyde of iron,	1.44
Lime,	.11
Magnesia,	trace.
Organic matter and water,	2.45
Potash,	.01
	99.39

It is perfectly similar to the sandy soils of Cape Fear. These examples of sandy soils are beyond the reach of the overflowings of the Roanoke, which always leave a rich sediment behind, and which is employed as a fertilizer, to a limited extent.

The marl is also too much charged with sand, in parts of the beds. The blue varieties gave the following composition :—

Sand,	65.60
Phosphate of lime, and oxyde of iron,	9.60
Carbonate of lime,	21.20
Magnesia,	trace.
Water and organic matter,	2.60
	99.20

Regarding the available matter in this marl as thirty per cent., it should not be ranked with the inferior varieties—though the sand amount to sixty-five per cent.

§ 37. The marl of Fishing Creek should not be passed over unnoticed. It consists of the three varieties, the red, blue, and consolidated marl. The blue has the following composition :—

Silex,	72.50
Phosphate of lime and oxyde of iron,	6.25
Carbonate of lime,	20.00
Organic matter and water,	1.25
	100.00

This blue variety underlies, or is beneath, the red or brown variety. The latter is composed of

Sand,	62.50
Phosphate of lime, and oxyde of iron,	10.00
Carbonate of lime,	25.60
Magnesia,	.11
Organic matter and water,	1.30
	99.51

Both varieties are more or less consolidated, indicating a favorable composition for agricultural purposes. The parts selected for analysis contained fewer shells than the general mass. They are small bivalves, so common in Wayne, at Goldsboro', and on the Tossnot, which is really of a better kind than the varieties containing larger and less decomposable fossils.

The shelly portion contains more lime, which is derived from the shells themselves; but less precipitate, which contains phosphate of lime. This variety gives the following composition:—

Sand,	15.00
Phosphate of lime and oxyde of iron,	3.75
Carbonate of lime,	80.00
Organic matter,	1·25
	100.00

The average quantity of lime is above fifty, taking the whole mass together.

Intervening between the two varieties, the blue and red, there is a more consolidated portion—a variety which answers to the appellation of stone marl—though it differs in its fossils from that of the Trent at Newbern, as well as from that at Wilmington. It gave me the following analysis:—

Sand,	17.50
Phosphate of lime and oxyde of iron and alumina,	7.50
Magnesia,	.12
Carbonate of lime,	72.12
Organic matter and water,	.50
	97.74

This variety exceeds the blue and red in the quantity of lime; and it appears that, as the sand diminishes and the lime is increased, there is an approach to the formation of a solid substance. The solidity and toughness, however, often depends upon a quantity of soluble silica, which, when present, forms an exceeding tough deposit, possessing many of the characteristics of a burr-stone. In this condition, the stone is unfit for agricultural purposes; but makes a durable stone for walls and fences. It is also an excellent fire-stone, and may be used for the backs of fire-places, though it is charged largely with lime.

§ 38. The foregoing samples of marl, derived from the same geological series, furnish, upon the whole, a uniformity of composition which was unexpected. It is true that a few of them contain an excess of sand, which I think due to accidental causes, and which does not belong to the deposit as a whole. It often happens that currents bear along sand in large quantities; and the position which the shell-fish are occupying, may receive, at times, large supplies of arenaceous matter. Sometimes, the lime has fallen to ten per cent.—the sand increasing in proportion. But the average proportion is thirty-three per cent.

The selections for analysis were not made with a view to obtain a maximum quantity of lime in the several beds; but rather an average. It will be seen hereafter, that the shell-marl differs materially in composition from a formation upon which it rests, or which is geologically older, and beneath it. There is, perhaps, in this older formation, more calcareous matter than is usually credited to it. The analyses which have been made have excluded the matter composing the shells, which it often contained in great abundance. They are very frequently entire and unchanged. In the shell-marl, the fossils, when small, decompose; and, though the matter in which they are imbedded is calcareous, still the fossils furnish, by disintegration, a large share of it, which is obtained by analysis.

The general aspect of the marl, as it lies in the beds, is quite the same. The thickness is variable, exceeding, in a few localities, fifteen feet: in others, it is less than three feet. There is, I believe, but one stratum which contains those fossils which have given it the appellation of middle ter.iary. I have not seen it divided into two, except the formation upon the Trent.

THE GREEN SAND—ITS COMPOSITION, ETC.

§ 39. Beneath the shell marl and belonging to an older formation, there are deposits which are of the age of the cretaceous rocks of Europe. I have referred to this formation before, and have stated that, as a fertilizer, it is superior to the shell marls which I have described. Its color is green, and when examined carefully, it is found to be composed of irregular particles of sand, which, when crusted upon paper, leave a greenish mark. The deposit is made up mainly of this matter. The most characteristic masses are upon the Cape Fear river, at a place called Black Rock, and upon the river banks near Wilmington. These beds belong to the same formations as those known in New Jersey as marl, and which are highly esteemed in that State, and which deserve all the praise which has been bestowed upon it.

The loss of the specimens collected for examination in the laboratory, rendered it impossible to furnish analyses of the green sand. I shall, therefore, give two or three analyses of the Delaware marl of the same age. It is important to know what it contains, and as there is a great uniformity of composition in all the beds both in New Jersey and Delaware, and probably in those of this State, its composition will be a guide to those who wish to use the same material upon the Cape Fear or wherever it occurs.

Its composition is represented by the following analyses:

Silica,	70.20	70.31	
Potassa,	6.10	6.51	
Protoxide of iron,	15.25	15 16	
Alumina,	3.14	2.63	
Water,	6.22	6.26	*J. Rodgers.*
	100.91	*J. Mansfield.*	

Professor J. C. Booth's Memoir of the geological survey of the State of Delaware, p. 71, 1841.

Another analysis, upon the same page as the foregoing, gave—

Silica,	49.30
Potassa,	9.16

Protoxyde of iron,	24.46
Alumina,	7.82
Water,	11.26
	100.00

The percentage of potash in the last analysis is nearer the average of the formation than the two preceding it. It is to potash that its fertilizing effects have been attributed. Most of this substance is entirely destitute of carbonate of lime. Still, if the fossil shells were intermingled with the materiel submitted to analysis, it would give a notable quantity of carbonate of lime. When the fossils are decomposed and intermixed with green sand, carbonate of lime is then found. It will be noticed in the foregoing analysis, that phosphate of lime does not appear in the list of substances which it contains; notwithstanding this, I believe it will always be found.

In a single specimen from Black Rock, consisting of the inside cast of a shell common to the formation, I found a remarkable amount of phosphate of lime. The substance examined differed in no respect from the general mass, being made up as usual of the grains of green sand, moulded to the inside of a cucullœa: the outside was removed. This specimen contained 52 per cent. of phosphate of lime. This was, no doubt, an accidental circumstance; by some cause or other animal matter had been preserved to that large amount. If this amount should be found in the casts of the shell so common in the formation, it will become an important fertilizer; not simply from the potash, but from the presence of an equally important element, phosphate of lime.

§ 40. The green sand in Delaware frequently consists of two portions—an upper and a lower. The lower is the one regarded as destitute of carbonate of lime. The upper is calcareous, and approaches in composition to the shell marl of this State.

Thus, the upper consists of—

Carbonate of lime,	18.6
Green sand,	33.
Sand,	35
Clay,	14
	100.6

Another analysis gave—

Carbonate of lime,	24.7
Green sand,	35
Sand,	31
Clay,	9
	99.7

§ 41. The foregoing furnishes the elements which may be expected in the lower deposits of the Cape Fear and Neuse. Black Rock is ten miles below Brown's landing. The green sand at this place is consolidated to the water's edge, and extends to an unknown depth beneath the water. There are ten or twelve feet above water, extending along the water's edge, against which boats may anchor. At this place, the marl is so accessible, that when the navigation of the river is practicable, boats may take in a cargo of it at a trifling expense. It is yet to be tried, and yet to be determined, how far this material will admit of transportation. Should the coal of Deep river find its way down the Cape Fear, and boats are returning empty or with light loads, it is not at all improbable that the green sand may be used in Chatham and Moore counties as a fertilizer. This is rendered still more probable, from the fact that fertilizers are rare upon the upper waters of this river, and the lands require something of this kind. Should phosphate of lime constitute an important element of this bed of green sand, it would bear transportation still farther, and admit of its use in the interior of the counties bordering upon the river and its branches.

The strata of rock at this place consist simply of the lower mass already described, and a bed of pebbles upon which reposes a thin bed of shell marl and sand. The green sand may be regarded as a continuous stratum, differing from the shell marl in this respect. At Mr. J. Sykes's, 9 miles below Black Rock, the formation appears again.

§ 42. On the Neuse, in the vicinity of Goldsboro', a formation appears of considerable extent, unlike the shell marl, and unlike, also, the green sand. Its fossils do not yet declare whether it is an upper mass of green sand or the lowest division of the tertiary. Fragments of an ammonite have been obtained from it, but the exogyra and belemnite, so characteristic of the green sand, have escaped detection up to this time. The formation is a consolidated marl or marl stone; of a light gray and a yellowish brown.

The following strata belong to a section near Col. Collier's plantation, to whom I am greatly indebted for the interest he manifested as well as in aiding the survey.

1. Green marly clay.
2. Marl, eleven feet, containing spine of echine, crabs' toes, &c.
3. Gray sandy clay.
4. Yellow clay intermixed with some gravel.
5. Sand.

The consolidated or stone marl lies upon a hill side. It is about six feet thick. It is granular, and might be employed as a building material, as well as lime for agricultural purposes. It is composed of—

Silex or sand,	39.20
Phosphate of lime and oxyde of iron,	1.60
Carbonate of lime,	55.20
Magnesia,	60
Potash,	trace.
Water and organic matter,	2.20
	98.80

The good effects of this marl appeared in its use upon an exhausted patch of land; as a consequence, it gave a fine growth of clover, which came in without sowing the seed.

§ 43. A stone marl at Wilmington, lying immediately upon the green sand, is composed of

Silex,	20.00
Phosphate of lime and oxyde of iron,	5.00
Magnesia,	.42
Carbonate of lime,	72.00
Organic matter and water,	2.00
	99.42

Intermixed with this consolidated marl are many fragments of coprolites, forming with it a very singular conglomerate. If it should prove extensive, it would form an excellent fertilizer. This specimen was obtained of Dr. Togno, at his experimental Vineyard.

Another specimen of stone marl gave the following results:

Silex,	27.40
Phosphate of lime and oxyde of iron,	1.40
Magnesia.	trace.
Carbonate of lime,	60.00
Water and organic matter,	11.60
	100.40

§ 44. The marl stones, the composition of which I have just given, require grinding to fit them for use. When fine, I should regard their composition superior as fertilizers to the shell marl. Careful burning in a kiln will fit them for use. In doing this, caution should be exercised not to expose them to a heat sufficient to fuse them. The greater effect which materials have upon the soil, when fine, might pay the additional expense required to bring them to the condition of a fine powder. The beds of stone marl are quite limited patches of ordinary marl, consolidated by the calcareous matter they contain. In mass, and exposed to the air, after removal from the quarry, they become hard; and in that condition they become quite good building stone, as well as a stone capable of resisting the heat of a fire without losing carbonic acid.

MODE IN WHICH MARL AND LIME PRODUCE THEIR EFFECTS.

§ 45. As the marls which have been described in the foregoing pages exert their influence upon vegetation, in part, by the carbonate of lime contained in them, it will not be out of place to speak of the mode, or modes, by which it is supposed to operate. Many theories have been proposed to account for the action of lime upon vegetation. It is even true that some maintain that its action is scarcely to be depended upon, or that it has any action at all.

A gentleman remarked, at a meeting convened for the purpose of discussing matters relating to agriculture, that he had tried lime upon a sandy soil, and it did no good; and he then tried it upon a clay soil, and there it did no good; and hence believed all that had been said upon the value of lime should be received with considerable deduction from the statements. Now, it so happened that both varieties of soil had been subjected to analysis, and it was proved that the sandy soil was quite deficient in organic matter, and that the clay soil was rich in lime, containing some three or four per cent. Now, it is conceded that one of the conditions required for the exhibition of favorable effects of lime is, that there should be organic matter; and, in the case of a soil already rich in an element, it is also proved that further additions of that element is not followed with visible results. Here were two cases, then, which had failed for want of judgment and knowledge in selecting the kind of soil upon which to apply a remedy, and an ignorance of the condition required to secure activity in the remedy itself. Undoubtedly, there are many disappointments of a similar kind, where experiments are tried, while ignorant of the condition necessary for the action of the remedy, or ignorant of the kind of soil upon which the experiment was tried.

I. So far as the plant is concerned, lime operates favorably upon vegetation by supplying an element necessary to the subsistence of the vegetable. Analysis of the ash of any plant gives an amount of lime in the state of a carbonate—not that a carbonate is the condition the lime is in when a part of the living plant. Any of the organic acids, combined with lime, become carbonates in burning. One of the uses of lime is to supply one kind of nutriment. The amount required by different parts of plants varies with the part. The outside integuments are richer in lime than the seed or woody part; and some plants require more lime than others. It is the food of the plant: and its use, so far as the plant is concerned, is all its use.

But, 2. Its use and effects in the soil are not so simple as has been stated with respect to the plant. As an alkaline earth, having a strong affinity for all acids, it combines with them, and forms a neutral salt; and this salt, being soluble, is taken up by the roots. We need scarcely speak of its power to decompose astringent salts—the proto-sulphate of iron, formed in those soils where pyrites exist, and which, when formed, are decomposed; and we may thus find sulphate of lime, gypsum, instead of sulphate of iron.

3. If, in soils abounding in peaty matter, organic acids are formed,—these will combine with lime, as already stated. But these acids being formed, and coming in contact with the matter of the same origin, act as preservatives; or they are termed antiseptic bodies, which prevent putrefaction. When neutralized, they cease to be antiseptics, or preservers, and the remainder of the unchanged matter goes speedily to decay. In this way, time promotes the decay of vegetable matter, and, at the same time, the salts formed become food for plants; and the salts formed are called organic salts of lime. So far, facts and theory support each other.

4. In the laboratory, lime acts as a solvent on silica; but it requires a high temperature. It is supposed by some that it dissolves it in the soil. But a more rational explanation is, that it decomposes silicates, consisting of silicate of alumina, potash, and soda; by which action, it brings silicate of potash into a soluble condition, forming also one of the elements necessary to the straw of wheat. These silicates are essential to the strength of the stems of grasses and cereals —when deficient, they are weak and lodge. There is, however, considerable obscurity on these points The affinity of lime for other bodies is strong, and it is rational to suppose that, in the soil, its action is not unlike that I have attempted to explain.

5. Resulting from its chemical action, there are physical effects upon the soil itself. No one need be deceived as to facts. Lime spread upon a stiff, well drained soil, makes it

light and loose. Now, this does not result from a mechanical mixture ; it comes from a prior chemical action, and the looseness is an effect due to that. It can be in no other way than that which results from decomposition.

6. Another effect of lime upon soils, especially when in combination with organic matter, is to give to light soils an increased retentiveness of water, and an increased power to absorb water. This I have proved by direct experiment. A merely pulverized limestone will not increase this power ; but marl, which is in a state of extreme subdivision, will, when it holds in combination organic matter. Hence, one of the effects of fine marl upon a loose sandy soil, is to give it a body, or a retentiveness of water. Marl put into a hill of growing potatoes becomes a fertilizer, while lime would destroy the vitality of the seed by absorbing its organic water. Marl will absorb ammonia, and thereby furnish fertilizing matter. From the foregoing views, if correct, it appears that lime acts *vitally*, *chemically* and *physically*. *Vital*, in being a constituent element of plants ; *chemical*, in its action upon silicates and organic matter ; *physical*, in imparting friability to argillaceous soils, and compactness to sandy ones.

7. The effects of quick lime and marl are not identical. They have, however, a common condition of the soil which is required for their action. I have already spoken of this condition : it is the presence of organic matter. I hold that organic matter is essential to the formation of an organic salt, both soluble and suitable to the nature of vegetating matter. It is proved by experiment, I believe, that thus far, the effect of marl and a sub-caustic lime may be identical. But sub-caustic and caustic limes are capable from their superior affinities for potash and soda, to do more in the line of chemical action than marl, especially upon argillaceous soils, where it may combine with, or as it is frequently called, liberate, potash ; and it is probable that marl may exert the same influence in a feeble degree. Marl, however, from its complex character, secures effects of a different kind. Referring to the composition of any of

the shell marls, it will be observed, that in addition to carbonate of lime, it also contains phosphate of lime. It will be observed, too, that some of the samples of marls are comparatively poor in carbonate of lime, but still they are represented as strong fertilizers; as producing in some instances four times the amount of corn and cotton as would have been grown without them. From these and other facts looking the same way, it is probable that the effect of marl is as much due to phosphate of lime as to the carbonate; and it may turn out that some of the argillaceous deposits which scarcely effervesce with acids, may yield a still larger amount of the phosphates than the marls, which give the largest quantity of lime. Hence, it will be important to look farther than the marls, and furnish chemists with all the varieties of deposits which are not essentially sandy. Potash, also, even the shell marls, should not be overlooked.

The beds associated with the marls proper, are various, and they open a field of research and experiment for the planter and chemist. Any of the argillaceous beds may be used in experiment. It is an important circumstance, that, although sand is predominant, in a large part of the lower country, there are beds of clay and marl at hand, which may be employed to correct this peculiarity.

§ 46. Some of the beds of marl, or those which are over, or beneath them, are highly pyritous, or contain sulphuret of iron. This substance, when it decomposes, under ordinary circumstances, forms an astringent salt, which is injurious to vegetation, certainly if in large quantities. But this substance is not to be thrown away and ejected into the streams. If its decomposition is effected in the midst of marl, and especially as a compost heap, it will give the farmer gypsum, a substance really important in all kinds of soils. Marl or lime is always the corrective of the astringent salts, such as copperas, or the sulphate of iron, or the sulpnate of alumina and potash, which may be furnished by stratas intermingled with pyrites. But small doses of sulphate of iron I believe to be useful:

hence, when pyrites is disseminated in a bed of marl, I should esteem it so much the more ; and, if found to be too strong, it is easily corrected or weakened, as it may be called, by intermixture of pure marl, or the pure carbonate.

There is a small circumstance worth a passing notice. The marls contain bones, as all know very well; and the laborers are in the habit of committing them to the waters, if upon the bank of the river or stream. As they are worth at least fifty cents per bushel, it will be economy to save them. If broken into small pieces, and dissolved with oil of vitriol, the best of fertilizers is made. When dry, one hundred pounds of bone require twenty-five pounds of acid for solution. This amount when well mixed with gypsum, or any neutral substance, to dry it and give it bulk, is sufficient for half an acre. But the bones, at present, are worth still more as objects of natural curiosity, and should be preserved for that purpose.

Sulphuric acid will cost three dollars per hundred weight ; it may be procured in the larger towns for two dollars and fifty cents per hundred weight. The object, in using oil of vitriol, is to make the bones soluble, by which immediate effects follow.

§ 47. The details relating to agriculture, which are spread out upon the foregoing pages, embrace those general facts which were observed and collected during a reconnoissance of several months. Although they may not contain all the information which many, and perhaps most, of the readers of this Report desire, and have reason to expect, still, I believe the way is prepared for pursuing inquiries to better advantage than before ; and which will conduce to a system of agriculture which shall be better adapted to the special condition of the soil, climate, productions, and labor employed in the part of the country and State under consideration.

Climate is a consideration which should not be disregarded. It must always modify and change, more or less, the systems of husbandry. If disregarded by the agriculturists, it will be at a cost and sacrifice of something valuable. If the soil is disregarded, it will be a loss in the crop; and neither can the character of the laborers, their capacities and strength, be left out of the account in making up our minds on the best systems of culture and of crops. It is of the utmost importance, to be correctly informed of the systems of culture, and the kind of crops most in use in foreign countries. But the entire system of means and ends should also be brought together in this information.

When we are told, for the first time, of the great value of the turnip crop in England, it at once wakes up incentives to adopt it here. When the system of agriculture, which is pursued in England or Flanders, is represented to us, there are strong temptations to adopt it, before we know the peculiar causes which have led to its general adoption in those countries. The kind of labor, the cost of labor, the intelligence and capacity of laborers, are important items of information, before we can safely decide upon the question of adoption or rejection. The culture of roots for feeding stock is carried to perfection in England, and it is a means of wealth. But, whether the means would crown the end, in this country, is quite questionable. England enjoys a humid climate; and vegetables, among which is corn, scarcely ripen at all. But, supposing it would ripen occasionally, can an uncertain crop become profitable? If corn would ripen every season, would the turnip and root system of husbandry prevail there? and may not roots, as food, suit better in the climate of England, than in the drier and less steady climate of America?

§ 48. I wish to inculcate and enforce the following doctrine: that every country, of sufficient extent to possess an individuality, favors certain productions; and, in order to carry these productions to the highest points of perfection to which they are susceptible, it may be done at a less cost in money and labor there than anywhere else;—labor and money, laid out when circumstances are favorable, than when they are unfavorable. But systems, often, have certain advantages which may be adopted; and an *eclectic* system, properly methodized and adapted to circumstances, may become, in every country, the best of systems. English husbandry, fully adopted in almost any part of our country, would fail in its ends; but there are points in it which may be adopted anywhere, and should be everywhere.

The application of mind, and the bestowment of thought, on husbandry and systematic agriculture, is what is most wanted. That, combined with facts and a knowledge of the principles involved in husbandry, will give direction to means and expenditure of labor and money to profitable results. The pursuit of agriculture, thus conducted, places the business on a level with the professions, and secures to the individual the same standing and influence which they have, and the mental ability to cope with them in the arena where mind alone controls and governs.

GEOLOGICAL POSITION AND RELATION OF THE TERTIARY BEDS—GEOLOGICAL TIME—MEANING AND ORIGIN OF THE TERMS, EOCENE, MIOCENE, ETC., AND USE OF THE WORD SYSTEM, ETC.

§ 49. The marls have been considered thus far as depositories of those elements which only interest the agriculturist :

which I have been speaking : it is their age, their relations to each other, and to other deposits, which may be in immediate proximity, or at a distance. Enquiries of this sort bring up many questions—questions which, at first sight, seem very easy to be disposed of, but which, on closer inspection, are found beset with many unexpected difficulties. However this may be, I remark, that these enquiries relating to the age of the strata, or to the system and series of rocks, and where they geologically belong, are not, in this region, easily determined. Opinions upon the question I have stated, have been expressed by the distinguished Professor Mitchell, of Chapel Hill, and by Mr. Conrad of Philadelphia, and others; and it is highly probable that these gentlemen are right in the main; but whether their views may not require a partial modification is yet to be seen. It is to be recollected that these strata of marl are scattered over an extended surface, and are disconnected with each other: they are isolated beds, lying in a belt of country, at least, four hundred miles from North to South, and two hundred miles in breadth, in its widest part; and that the slight disturbance which this belt has undergone, since the era of the green sand, is too slight to show uncomformability. We are therefore obliged to settle the question, by reference to the fossils of the beds, and the animals now living in the Atlantic; relying upon the determination of the ratio of the dead to the living species.

§ 50. But, before I proceed to speak of the age of these deposits, it seems necessary to make some preliminary explanation. When we speak of age in geology, the idea of time is involved, and the question might well be put, how long a time, or how many years ago is it, since these strata of marl were deposited.

Time, when it enters as an element in geological reasoning, is not expressed absolutely, but relatively. Time, considered as an element in human affairs, is both absolute and relative. It is absolute in all our calculations, because it has its units. A rotation of the earth upon its axis is the unit; and its revolution round the sun is three hundred and sixty-five and one-fourth of this unit. As it is the movements of stars which give us the unit, it is called sidereal time ; and all events which have transpired since man was created, is measured by this unit, obtained from the observations upon their motion. We have data, therefore, for absolute time. In the history of man, then, there are two facts not known in geological time,—a *unit* and a *starting point*.

Geologists have sought for a unit, but have failed. Sir Charles Lyell has counted the layers in the sediments of the Mississippi, and measured the suspended mud in its waters, which it annually brings to the sea. Rationally, it carried him back forty thousand years, since they began to be formed ; but the unit was still problematical. He has counted the steps of the Niagara, as it has receded from a lower to an upper lake; but the steps are unequal. It can give us no unit in its march. Unlike the earth and planets, which rotate in equal time, or which revolve in great circles, and return to the starting points in equal time, the geological movements of all kinds are unequal, and their perturbations so great that they give us no unit to measure geological revolutions by. For us, space too is a unit, and it gives us a measure for time ; so that time is space, and space is time ; but geology cannot convert space into time, nor time into space.

§ 51. But geologists have sought for units in life ; the search was foiled. Even man, whose life is most exalted, gives no true measure, no unit, either in his individuality, or in the sum and aggregate of his generations. Here, we can scarcely refuse to inquire, why geology gives us no unit by which to count the years of the earth? But then the enquiry is futile : and we can only say, that it can have no final cause—it can have no practical use. But the determination of relative time is of immense use—it is practical.

It has a practical application to the relations of our corn fields, our iron and our marls. Time then, in geology, is only relative, as it has no unit. If we take geological movements in upward space, or horizontal space, we find these movements have taken unequal times; and, in equal times, the movements have been unequal. Vertical elevations may amount to three feet in a century, and the sediments may accumulate thirty feet in thickness in five centuries ; and yet, when centuries are compared, both the sediment and the vertical rise are unequal in equal times. The chronology, then, of the earth, is computed relatively: time, which measures its history, is divided into unequal periods, and those periods have visible representations in the sediments of its crust. Their super-position upon one another, exhibits their relationship to the eye, and the contemporaniety is proved by the oneness of their products, and the sameness of their representatives of a former life—the remains of the dead entombed in the rock, whose formation was in progress when they were the sole possessors of earth.

§ 52. Taking the visible representation alluded to, as time, in a geological sense, our computations are made possible by breaks in the series. They are not, however, like the continous and regular clicks of a chronometer, which divide out the time into parts of a unit; the breaks not only separate physically the visible representations, but put to a stand-still the currents then bearing onward humbler forms of life, and their burthens of matter. The break is an upward movement, and it marks the end of a dynasty. Strange as it may seem, the earth's crust is marked by sychronous breaks, almost universal. Breaks which pursue the directions of mountain chains are sychronous; and parallel mountains are formed by sychronous breaks and elevations. After all, there is the semblance of law in the movements and breaks, which have seared the earth's crust, and brought to an end the existing order of things. The area of sediments is changed—the direction of the rivers is changed—the life, as it is represented in species, is extinguished and changed. But, again,—after the turmoil attendant upon change has passed away, life begins to be lit up in new abodes ; and as each break is followed by new physical relations, the life which appears is modified to fit the change. Species are, therefore, new; typical forms remain; and the great types upon which life, and its various forms, were to be represented, are preserved. The great scheme upon which life and its phenomena were to be displayed, has never been broken or departed from. Each break, and the new sediments which follow, indicate a new period, having its beginning in the lowest of the strata; and, as these new sediments are going on, life is still ebbing and flowing, and the individuals which are dying are entombed in the accumulating wastes of the older continents. The breaks, then, mark the outgoing and the incoming of new systems. The space spanned by two breaks is one of the unequal periods in the earth's history and progress. The distinguished French Geologist computes no less than twenty-seven breaks, which have destroyed the existing and living Saurians; each break approaching to universality in their effects. The earth's crust is a sepulchre. Its sediments, which are ten miles thick, are full of the relics of plants and animals.

§ 53. As we are unable then, to compute in years, when the present order of things began, or when the lower orders of animals first appeared upon the earth, Sir Charles Lyell has proposed to express the simple relations of the past to the present, by words adapted to that purpose. Observing, for example, that the chalk beds are destitute of any species of animals and plants, which now exist, and that the succeeding beds contain them, he takes his starting point from the latter, and attempts to express their relations to the present, by terms of comparison. The mode of proceeding, in order to determine the comparative expression was, first to ascertain the whole number of species of fossils in the beds, and then ascertain how many are living in the seas of the present day. In 1830, the number of species known and described in the beds next above the chalk, was 1238. The number, since that period, has increased to 5000 at least.

Of this large number, three and a half to four per cent., are living now. The small number, then, ascertained as the surviving species, which have resisted the change and revolutions of the globe, indicate an approach to the conditions which now exist upon the earth's surface, and it has been regarded therefore, as *the dawn of the present*. The word *Eocene*, which means literally the *dawn*, is applied to those beds, or that formation which has preserved from three and a half to four per cent. of the species which now live. It is applied to the lower beds of the tertiary system. The beds which are typical of this part of the system, underlie Paris and London, and, being basin-shaped, they have been called the Paris and London basins. The former was the field in which the celebrated Cuvier labored, whose name and labors have conferred honors upon the French nation. But time moves on; and the next step in the succeeding series shows an advance; they contain seventeen per cent. of species or kinds, which live in our present seas. The strata containing this per centage, repose directly upon the *Eocene*. As the proportion of the dead to the living, bears still a small ratio to the living, or is a minor proportion still, of the dead to the living, the formation is called *Miocene*. The rocks which succeed and rest upon the latter, are called *Pliocene*—meaning that the proportion of the living is more than the dead. When the per centage of living species is thirty-five and fifty per cent., the term is qualified, and is called either older or lower Pliocene; and, as the per centage still increases, in the succeeding deposits, and rises as high as ninety-five, the portion of the series is denominated *newer* or *upper* Pleiocene. Certain beds which compose the Sub-Appenine hills, and which are very thick, and attain an elevation of fifteen hundred feet, belong to the latter. It would be interesting here, to illustrate the slow accumulations of those beds, for the purpose of imparting a correct view of the great length of time which passed while the beds were being deposited; but, I can only state the fact; and then proceed to say, that ever since the full dawn of the present, beds of immense thickness have accumulated in a slow manner, filled with marine and terrestrial animals, most of which are identical with the present. But it is found on careful exploration, that even in these modern deposits, one or more species out of a hundred, have become extinct; and that some of the living ones have undergone a slight change; some which are quite small now, were formerly large. Causes then still operate which produce a change in a minor degree, and which alter those characters which are the most susceptible of the influence of physical agents. Those modern deposits, which furnish evidences of slight changes, are called *Post-pliocene.*

§ 54. The names, the origin, meaning and application of which I have given, are subordinate beds of the tertiary system. It has been separated from the older upon which it reposed, because all of its beds or sub-divisions contain some few species which are not extinct; the lowest or oldest furnishing the smallest proportion, yet constant, of all the others which succeed; the newest, the largest proportion; and, as succession is proved by the position of the bed, one above the other, so we may infer that there has been, or was progress also. This is shown, by facts which have not been stated, viz: the increase in number and rise in rank of the terrestrial animals, which only began to figure in the tertiaries. This progress is indicated by the increase of those kinds of animals whose physical constitutions approximate very closely to man.

§ 55. Notwithstanding the plausibility of the arrangement, and sub divisions of the tertiary beds and the euphony of the terms, Eocene and Miocene, &c., applied to them respectively, it cannot fail to impress us that it is artificial and arbitrary. For this reason, probably, we find it difficult to apply it to the tertiaries of the Atlantic coast. It seems to be more applicable to European tertiaries; and, it is very possible, that it is because they are better known than ours. It is not impossible to apply the scheme to our formations, when considered by themselves; but, we do not succeed as yet in fixing the relations which our tertiaries bear to those of Europe. It is, however, a popular scheme, and has been adopted by our Geologists and writers; and hence, it will be difficult to replace it by any other, even if it should be founded upon characters more natural and less arbitrary.

§ 56. In this connexion, it will not perhaps be amiss, to explain the word *system* more fully than I have yet; though its geological meaning will not probably be misapplied. A geological system is a series of rocks, forming a subordinate part of the earth's crust, which were formed and consolidated during a period when the physical conditions were the same, or nearly so. As there are many systems, it is implied in this definition, that the earth's crust has been subject to change; and that the periods are marked, in the beginning and ending of these periods, by certain changes which they have undergone. We know thein comings and out-goings of periods, by phenomena visible in the rocks; they are tablets containing the records; and it is a remarkable fact, that the records are not confined to physical changes; they also extend to, and embrace, those which concern the species of plants and animals of each.

§ 57. It is agreeable to observation, that a vertical movement of the rocks is accompanied by a decided change in the kinds of animals and plants which had previously lived; they becoming mostly extinct, while their successors will differ from them in kind. Now, our definition of a system, and the remarks following it, seem to make the vegetable and animal kingdoms subordinate to the physical or inorganic kingdom; in other words, that both are controlled by physical forces; and that these forces being modified in intensity, are the causes which are instrumental in destroying the existing kinds ; and, also, of favoring certain special kinds in the succeeding system ; or, that the latter are consequents of the antecedents referred to.

The changes of species, resulting, as is supposed, from movements of the earth's crust, will not appear strange, when we are informed that it at once involves a change in climate ; a change in the humidity and dryness of the atmosphere ; a change as to heat and cold ; all of which exert, upon all kinds of living matter, important effects. Life requires the presence of certain elements, as oxygen, carbon, etc., and also, an apparatus upon which the elements act, as the lungs, stomach, etc. These must remain, in some form, whatever change takes place.—But life may continue under an infinity of modification of apparatus, and we may suppose, that the modifications of this kind are intended to adapt species to the minor conditions of life and their fluctuations arising from changes in the earth's crust. All the important and controlling elements still in force species of both kingdoms live on—the minor changes not affecting life in the abstract, but only the form of the apparatus. Minor fluctuations affect only external forms of apparatus ; the different species are characterized by these modified external forms. The origin of species is consistent with changes in the position of the earth's crust, and its relations to water, which modify the condition of the atmosphere, the temperature of the ocean, and its depth. These conspire to modify the causes which fit the media, in which organic bodies live and move, to execute the functions of life on the best and easiest terms.

Sir Charles has proved that the temperature and humidity of any given place are modified by proximity to the ocean or to waters, and by height; and the creation of new species is required to fit them for those alterations consequent on the change in vertical height.

§ 58. Vertical movements are indicated by breaks in the series, which for the most part run in certain directions for great distances. The Appallachian chain of mountains, the Blue Ridge, are instances which show a vertical movement at some former period, accompanied with breaks, or fractures of the earth's crust. Both the range and fractures are parallel, and extend through many degrees of latitude. But many chains traverse the earth's crust; all accompanied with breaks of the strata. These pursue different directions; and it seems that there is a law involved in these facts: for example: those chains which run in the same direction were elevated at the same time; the force beneath, which produced the breaks, or whatever that form may have been, operated simultaneously in one direction. The forces have been operative continuously, or in paroxysms. When in paroxysms, the period of activity is short, followed with long intervals of repose; when operative continuously, the period of activity is long. The paroxysmal modify the minor conditions of life speedily; while the slow scarcely produce distinguishable effects in the historical era, and not at all in the life of man.

I see no objections to the foregoing views, the subjection of species or life to physical forces and elements; for it does not imply that final causes, respecting those forces, had no reference to life. They precede life in the order of time; but the entire machinery may have been devised, and put in operation, with the sole reference to life which was to come, and to give dignity and importance to those forces, and to the arena upon which life, and its attributes, were to be displayed.

BEDS AND STRATA BELONGING TO THE TERTIARY OF THE LOWER COUNTIES.

§ 60. The periods represented by the shell marl beds, or the fossiliferous deposits being included, and except the green sand, are:—1st. Post Pliocene; 2d. Pliocene; 3d. Miocene; 4th Eocene; beneath which, repose the green sands, which belong to the system known as the cretaceous, or chalk or cretaceous system.

1. Post Pliocenes.—Under this head, I include all those beds ranging along the sea margin of the Atlantic slope. They consist of isolated beds of the recent oysters, containing, also, the more common molusca of the coast; but the most common by far, are the common oyster, (Ostrea edulis,) and common clam, (Venus mercenaria.) They are known as oyster beds. They are usually regarded as collections and accumulations of shells by the Indian tribes, formerly occupying the coast. This view may be sustained in some instances,—but it often happens, that the accumulation or beds are too extensive, and contain, withal, a mixture more accordant with that view which ascribes them to the operations of nature—which regards them as beds made up of individuals which grew upon the spots where we now find them, and which have suffered a vertical movement which has raised them above the level of the ocean. The beds occupy the more elevated points, or the rounded eminences; which, while submerged, were the summits of low hills, below the agitations of the waves. It is needless now to dwell upon the characteristics of these beds. They are clearly, and without doubt, to be refered to the Post Pliocene; and, though they may be regarded as possessing the characters of the present species now occupying the coast, still, it is not difficult to find in-

dividuals which are very large among the common-sized ones; and there are even beds, all the individuals of which are larger than those now living upon the coast. Near Newbern, beds of very large oysters occur, especially on the plantation of Mr. Pasteur.

It is highly probable, that different beds were elevated at different times; and hence those farther inland, all things being equal, are older than those immediately upon the beach.

§ 61. It may, at first, appear a startling statement, that our coast is subject to variations of level. Of the truth of this view, however, there can be no doubt;—so numerous are the phenomena indicating vertical movements, that it has, at many points, become common belief.

A subject of so much interest is worthy of farther consideration. I spoke of a vertical movement, as if this comprehended all the facts of the case. This is not so. The facts go to show that there are oscillations—that there have been both upward and downward movements. The upward movement is proved by the existence of raised beaches of sand and shingle, and beds of oysters and clams 50 or 75 feet below low water mark. The downward movement is proved by the remains of submerged forests, consisting mostly of the stump of the present pine which grows along the coast. This downward movement is far from being confined to a few miles of coast;—it affects both Sounds their entire length. The stumps are now to be seen in shore: some at high water mark; others at low water: and more standing far in, and constantly submerged. So common are these old stumps, that the fishermen are obliged to incur a great expense in clearing off these submerged stumps from the bottom of both Sounds—as there is no fishing ground which is free of them. At the present, it is perhaps impossible to determine which way the Sounds are moving. I am of opinion that there is not a uniformity in this respect;—that, while parts may be actually sinking, others are rising, or stationary. The street along the water's edge, in Beaufort, was laid out many years ago, and marked by cannon, set as posts, deeply in the sand.—One portion of this street is constantly beneath water now, as if there had been a change of level since the street was laid out.

§ 62. While upon the subject of change of level of the great Sounds, Albemarle and Pamlico, it will not be foreign to the subject, to notice the change which has taken place in the saltness of the waters of the Sounds. This is not supposed to arise from a change of level—a subsidence—though subsiding is not to be left out of view. It is attributed mainly to the opening of new inlets, by which the ocean's waters have a freer communication with the sounds. The freshness of these waters had become such, that marine shell-fish had died out; but since the opening of the new inlet, the waters are in the act of being peopled again with marine animals. We cannot but notice, in these facts, what has taken place in other parts of the world, and other times than our own—where many alternations of fresh and salt water had occurred, each containing the fossils peculiar to that state.

§ 63. But to return to the Post Pliocene deposits. I remark, that they do not contain the bones of extinct land quadrupeds, such as the mastodon, elephant, horse, &c.—that is, none have as yet been discovered in them, though sought for. It goes to prove that these quadrupeds had already become extinct, prior to their formation; or, I may say, the evidence leans strongly that way, when all the facts are told. The coast oyster banks are the latest formations—the newest:—and probably their elevation or reclamation from the oceanic waters has been effected in times only just anterior to the historical period.

§ 64. There is what may be called a deposit upon the banks of the Roanoke, which is worthy of notice, for this reason: it marks distinctly the difference of deposits, which have been made in geological time, from those made in absolute time. This deposit consists of fresh water shells, and contains along with them the bones of the turtle, alligator, turkey, dog, deer, and those of man; together with the rude utensils common to the savage state. It is a bank, one fourth of a mile long, and twelve rods wide, and raised eight feet above the adjacent plain. The part abounding in these remains, is about four feet thick. This notice of a formation is important only as illustrative of the distinction between the ancient beds and the modern, containing the bones of man, together with his implements of war, and his apparatus for cooking. It is characteristic of all deposits, the world over, which contain the remains of man, blended with the remains of animals, all of which are now living; all going to prove that man has not been an inhabitant of the earth only for a short period; inasmuch, too, as his remains are found in none of the formations containing extinct species of either land or marine animals.

§ 65. The Post Pliocene beds are co-extensive with the Atlantic coast. They are naked beds; or with the slight covering of vegetable mould which rarely exceeds eighteen inches—usually less I have seen large trees growing over the beds; but, in many places, they are naked wastes—as at Nagshead. These wastes are often exposed to the furious Northeast winds which sweep over the sands and hills, and bear them inland It will be seen, then, that a beach is not wholly raised by vertical movements effected by a subterranean force. Upon the Carolina coast, the breakers carry forward the sand; and, when they flow up the inclined plane, the sand is spread out with great evenness, and then left there by the receding or retiring wave.

The wind, when strong, sweeps overt he dry and loose sand and bears it still farther: when it becomes drifting sand. The coast gains more than it loses; and not only are the sands brought up, but pieces of wrecks of vessels, iron bolts, spikes, etc.; and even silver coins, from the sunken wrecks are sometimes found. A spear or fishing tackle, which is lost overboard some 20 or 30 rods out, will be sure to be found upon the beach in two or three days. This, perhaps will not happen on all shores; but those formed and acted upon, like those at Nagshead, favor such a result.

§ 66. Pliocene.—Anterior to the post pliocene, the beds which were deposited, whether in small basins, or in the form of irregular belts skirting the seashore, or in caverns where terrestrial remains of extinct quadrupeds are found have received the name *pliocene.* The pliocene admits of a sub-division: and has been designated, according to position: the oldest and inferior, as it contains a larger proportion of extinct species than living ones, is called *older pliocene*:—the superior, which contains fewer extinct than living ones, is the *newer pliocene.*

Pliocene beds are not unfrequent in North-Carolina;—but the beds which I now regard as pliocene, are not fully determined as such: as the evident intermixture of fossils of pre-existing formations, and the present uncertainty of the species now living upon the coast, renders the application of the rule of per centage uncertain: And I may go farther, and question whether it is not impracticable to draw lines of demarkation between the pliocene and the miocene strata, for the same reasons.

Following, however, the guides which have preceded me, I shall refer certain beds to both divisions of the tertiary—the pliocene and miocene—without attempting, however, to show to which division of the pliocene any of the marl beds belong.

§ 67. To the *pliocene*, I refer certain beds near Newbern—those upon the plantation of Mr. Donnell. To the *miocene*, I refer the beds upon the Tossnot and Little Contentney Creeks, in the upper part of the valley of the Neuse, and between the Neuse and Tau rivers.

In the beds upon the Tossnot and Little Contentney, I found the ear bones of whales, (*cetacea*,) probably true whales, of the family *Balœnidœ*, and their vertebræ ; and also bones of the mastodon; and a species of *Orbicula*, differing from the only recent one I know. Those of Fishing Creek, a tributary of the Tau, may, also, be referred, perhaps, to the older pliocene; but not certainly. Those of Fishing Creek contain pectens, which are referred to the miocene by Mr. Conrad. The investigation requires to be continued.

To the miocene, also, I refer the beds at Rocky Mount, Tauboro', and Goldsboro'. The bones of vertebrated animals are found in all of the beds at the localities, particularly those belonging to the mastodon. We are obliged to refer the *mastodon giganticum* to the *miocene* :—the tooth, or portion of one I procured at Tossnot is not the tooth of the *N. Augustidens* :—but we have associated, with these bones, the large *pectens*, P. Madisonius, &c.

The section at Tauboro' exhibits the following strata :—

1. Sand, at the water's edge.
2. Clay, containing lignite.
3. The shell marl, with abundance of pectens, *P. Madisonius*, eight feet, contains the fossils; and three or four feet of clay, without fossils.
4. Sand and gravel intermixed.

The marl contains coprolites. Rocky Mount furnishes a similar section, with similar fossils, resting on granite.

Near Newbern, the beds which have been referred to as *pliocene,* contain fulgur canaliculatus, and fusus quadricostatus, (*miocene fossils,*) astarte natica, fissurella, calyptrea, pectusculus, &c.; the large beds of the other pectens being absent. The fossils of the Newbern beds, already spoken of, on the plantatior of the Hon. Mr. Donnell, contain a large number of sheils which I am unable to distinguish from those of the coast. Mr. Donnell's beds are white, loose beds with crumbling shells, more or less chalky, in consequence of being above water. The opening is recent—the bottom had not been reached. A fragment of a bone of the *mastodon* was found also in this place.

We can scarcely avoid comparing this marl, with its accompanying fossils, with the crag of Suffolk. The remarkable display of sands, gray, red and brown, embraced in these beds, assimilate the entire formation of this age upon our coast to the crags of Suffolk, (England,) and the fahluns of Touraine, in France

§ 68 The beds of marl, upon the Cape-Fear, at Elizabeth; at Walker's Bluff; those of Messrs. Lassaigne, Cromarty, and others, have also been referred to the miocene period. At Elizabeth, the strata are various; consisting of sand, clay, with light sandstone, marl, &c. &c., terminating with colored sands, as follows:

1. Sands at the bottom of the cliff; gray and thin seams of clay, and some lignite.
2. Bluish and sandy marl, pyritous.
3. Thin beds of coprolites, pebbles mixed with shells and sand.
4. Consolidated sand, with fossils, *arca.*
5. Marl, three or four feet thick.
6. Ferruginous sand, with diverse stratification.
7. Blue clay.

8. Surface sands, of various colors, twenty-five feet.
9. Between the clay and surface, same red conglomerate of Fayetteville. It thins out before it reaches Elizabeth, being the surface sand, which is very thick in the pine forests, and rests upon tno thin stratum of blue clay.

It is probable that the marl rests upon the upper eocene. The changes from the sands below, the rolled pebbles and coprolites at the bottom of the marl, indicate a change, and show the propriety of separating the upper from the lower beds of the bluff or bank. The beds of Elizabeth, Bladen County, abound in teeth of placoid fishes—a single tooth of chacharodon megalodon, saurian teeth, and a middle portion of thigh bone of a large saurian But, as the teeth and bones are more abundant among the rounded stone, it looks highly probable that they may have been derived from some older rock.

§ 69. At Brown's Landing, the bed of marl in the bank contains fossils of the same kind as at Elizabeth and Walker's Bluff; and also contains many individuals of the exogyra costala, a fossil regarded as characteristic of the green sand, (*cretaceous.*) These individuals are in a fine state of preservation—some large, and others small—but none of them have been rolled on a beach. Both valves are together, and the fossil is in a perfect condition. No belemnite, an almost constant associate, has as yet been found at this place. Notwithstanding the presence of the exogyra, I am disposed to regard this bed, as well as those above, as miocene; on the ground that these beds are derived from the green sand. This view is supported by the fact, that one-fourth of the bed is made up of the particles of this cretaceous rock. In the same position I place the marl of Mr. McDowell, one mile from Brown's Landing, the marl of Miss Andress, where the exogyra abounds, both young and old, associated with an *oliva*, identical, I believe, with the one living on our coast, near the Fort at Smithville.

§ 70. Whatever may be the result of inquiries respecting the age of the shell marl deposits, it is plain that the only mode by which satisfactory results can be reached, is, by a copious collection from all the beds ; and, from the coast, of all living species. This should be undertaken ;—for the questions are involved in obscurity and doubt ; and although this course does not appear to advance the economical objects of the Survey, still, it usually turns out, that what appears, at first, only a scientific interest, does, in the end, promote, also, the practical application of the facts discovered, or already known.

The majority of the shell marls of North-Carolina are referred to the miocene period, by Professor Mitchell—who is sustained by Mr. Conrad, of Philadelphia. Those of this subdivision of the tertiary, which are far inland, as those at Elizabeth, contain very few, if any, of the molusca of the green sand, and, perhaps, very few of the eocene; while farther below, as at Brown's Landing, the lower fossils are very common, and they appear as much at home there as any of their associates. It is probable, then, that this occurrence is due to the proximity of the beds to which they belong. It should be observed, that these deposits of shell marl are in banks—which does not favor the view generally entertained that they reposed in and upon the strata upon which they lived and died. They seem, rather, to be beds formed by the action of waves, which have piled them together in great disorder—though they are not water worn.

§ 71 Eocene.—The lowest, or oldest bed, which I am able to refer to this formation, consists of pebbles rounded by attrition. They are beds from fifteen to twenty feet thick; and, at their western outcrop, form rounded hills, as at and in the vicinity of Carthage, in Moore County.

Similar beds and eminences traverse the State. In the vallies of the Roanoke and Dan rivers, they extend beyond Leaksville. They overlap the pyrocrystalline rocks, the granites, and gold slates, lying beyond the fossiliferous beds, which succeed them in the ascending order.

The extension, eastwardly, towards the coast, cannot be marked or determined very satisfactorily. Thin beds of rounded pebbles are known beneath the fossiliferous beds; but nothing interesting has been elicited concerning them. The pebbles are pure quartz; derived from the quartz veins of the gold slates In many places, the pebbles are cemented together by iron;—the coarser and finer sands are also cemented, forming a pudding stone. These cemented masses have taken various imitative forms:—as tubes, balls, cups, &c. The quantity of iron investing the clay and sand is sufficiently large to pay for extracting it for working. It often furnishes good limonite. The origin of the cemented beds must be due to ferruginous springs, which have ceased to flow; but which bring up the carbonated oxyde, and flowing subsequently through and over the beds, have filled the interstices with ferruginous matter. This, adhering strongly to the stones and sand, by this means has formed, finally, a pudding stone, by cementation.* The ancient beds, which consist of rounded stones and coarse gravel, with only obscure lines of stratification, are called shingle beds. They mark the beginning of a new order of things; and, hence, are important, as a means for defining the boundaries of systems or formations.

The term *pudding stone* has long been in use; and I apply the word to cemented masses formed above water; while the term *conglomerate* is applied to those cemented masses, or cohering pebbles, which have been formed beneath the water.

Thin beds should never be regarded as similar, in their origin, to norther drift, or transported rocks, or transported gravels and sand ; at least, in the mode in which materials, which have the same form, at the North, have been transported. There is not a boulder or a drift bed in North-Carolina. The masses which have been moved in this and other Southern States have been by means of rivers and oceanic waves—those means which exist now, and are in operation under our eyes.

But, to return to the ancient shingle beaches—I observe that they form the outer rim of all the tertiary deposits—a rim which, it is true, may be interrupted in places; but they range in a line, and cross the State to the westward of the first fall of the principal rivers which drain the Atlantic slope.

§ 72. The beds which succeed the former, are clays and sands of a greenish color, cherty clays and marls, together with interrupted beds of consolidated marl. The latter assumes the condition of a porous rock, sufficiently hard to form a building stone. It is an impure limestone—carbonate of lime forms about three-fourths of the rock. Soft marl underlies the rock.

The thickness of all the beds which I now regard as *eocine*, is not well determined—the limestone, or upper part of it, is only five, six and perhaps ten feet, in some places. At Col. Collier's plantation, near Goldsboro', it is only five or six feet. The lithological characters vary very much at different points; and sandy beds are replaced by cherty ones, or the cherty clays and limestone.

At Wilmington the rock is extremely tough and hard—though porous It is highly silicious. Beneath this, is a softer portion. made up of carbonate of lime, which is intermixed with broken and rolled coprolites; forming a conglomerate. This portion of it is highly valuable as a fertilizer, and has been employed as such by Dr. Togno, at his vineyard near Wilmington. This gentleman's enterprise is one of great importance; and the results of his experiments will be, to throw light, not only on the marls as fertilizers, but upon the vine and other fruits which will bear cultivation in this State.

§ 73. At Wilmington, the fossils consist of scutillœ royersi; one or two species of echinodens; teeth of the genus charcharodon sulcidens, galeocerda pristodontus, lamna elegans, &c.—the latter of which are by far the most numerous. The teeth of sharks, which are so numerous, and of which I procured many species, lie in greater numbers at the bottom of the higher and newer deposits. They should be regarded here as characteristic only of the oldest formations in which they are found. Their hardness and form favors very much this removal from the older to the newer rocks, wherever the latter derive a portion of their materials from the former.

§ 74. Some of the eocene, as well as the miocene, beds, contain numerous bones;—these are generally broken—even the thickest are broken to pieces, some six or eight inches in length;—thus, the femur of a saurian, one and a half inches in diameter, was broken into three pieces—the ribs of a whale into pieces about eight inches long. In these fractures, we observe the spicula of bone, still sharp, and never rounded or worn. These fragments are found embedded with delicate unbroken shells—a fact which throws considerable obscurity upon the causes which have broken them;—for it does not appear, from any phenomena, that these beds are subjected to a disturbance, or to a force, since deposited in the beds in which they now repose, which could possibly break such strong bones into such short pieces; especially when delicate shells are presented entire. And it does not appear that they have been subjected to attrition,—to the action of waves or stones.

§ 75. It has been generally supposed, that the bones of the whale, and mastodon, are found in the superficial coverings,—in those beds which are of the same age with those at the North.

Now, the mastodon is found in the fresh water marl of New York, and other localities; or in beds which repose upon that stratum called *drift;* and which is entirely wanting here. But, in North Carolina, they are found in the miocene, or older pliocene. The species of mastodon seems to be the one which is found in New York.

The question comes up, are these Southern beds of the same age with the fresh water marl beds of the North?

The latter are regarded as *post pliocene.* The bones of the horse and deer are also found in the same beds with the mastodon, and belong to the same age. I procured a grinder of the horse, at Greenville, in the sandy strata, just above the miocene marl. All the extinct deer and oxen, in New York, are found, also, in the fresh water marls; associated with the fresh water shells; the species of which are now living in our lakes and ponds;—and yet, the quadrupeds, in both formations—one in the South, the other in the North—are all entirely extinct.

If it should turn out, that the mastodon in the North Carolina marls is a species, specifically different from that of the fresh water marls of the North, the case would not involve the question of comparative age of the beds in which they occur. If the species are the same, it is difficult to reconcile the fact with the present views of Geologists upon that subject; the age of deposits as deduced from their fossils.

§ 76. It is impossible to define at this time, the limits of the eocene beds It is difficult to sub-divide the formation clearly, though it appears, that, it admits of the same division, as in Alabama and Mississippi—the cherty portion beneath, and the consolidated marl, or the marl stone above. But this part furnishes very few fossils. There is still a mass, quite sandy, similar in outward appearance to the green sand, which forms a feature, which should not be overlooked, in making the natural division of the strata. For agricultural purposes, the best beds are the marl stones; or those immediately beneath, which are sprinkled with fragments of coprolites. When first removed from their beds, they are soft, and easily crushed. When they have been exposed to the atmosphere, and have lost their water, they become hard, and crush with difficulty. They may perhaps answer a good purpose in building and construction.

THE GREEN SAND AND ITS GEOLOGICAL RELATION.

§ 77. A substance which is well known to every person, in the form and under the name of chalk, is a rock which has given the name to the system, of which the green sand is a member. But the chalk itself does not exist in the United States. And the system, to which the green sand belongs, has its principal representative, in the inferior or oldest deposits. While no true chalk is found in the United States, possessing the characters of the writing chalk of Europe, there are still deposits, which nearly correspond in age, with it. This view is supported by the numerous fossils, which certain beds contain, identical with those of the chalk of Europe.

This formation* is very extensive in the United States. From New-Jersey, South, to Alabama, it is one of the most continuous deposits.

In North-Carolina, it is concealed by the soil, except in favored positions. Upon the Cape-Fear, and its tributaries, it is probably better exposed, and more accessible, than at any point, known to me, farther North.

It consists of a series of beds, mostly sandy, alternating with a few inconsiderable beds of argillaceous sand, colored with chloritine matter. There are beds along the green sandy formation, which are calcareous, and which I now class with the eocene; but which may hereafter furnish facts, which will place those calcareous beds in the cretaceous system. These calcareous beds, however, are destitute of many, if not all, the characteristic fossils of the green sand, or cretaceous system, unless the single species of ammonite should prove to be one of them; though, I believe it will constitute a new species.

§ 78. The green sand beds are not distinguishable from the eocene, by the presenceof the green matter, which has given it this name, as it is, also, quite common in the overlying beds. This formation is beneath the shell marl, which contains those large scollops, and generally with beds, which are composed of aggregations of shells, closely resembling those living bordering on the Atlantic. The beds, of green sand may, however, be known by the presence of a cylindrical fossil, of a yellowish color, and which is 3 inches or more in length, and tapering to a conical point at one end. When unbroken, it has a conical cavity at the other end. It is called by some a thunder bolt Its name is *belemnite* It occurs upon the Cape Fear, at Sykes' landing, some distance below Black Rock, and it should be found at the latter place also. Another fossil which is very common is the exogyra costata. It is something like a thick rounded oyster shell. This is abundant at Black Rock.

I do not attempt, at this time, to speak of the extent or distribution of this formation or its thickness. But, I wish it should be known, that it is an important rock; that it is one of the best fertilizers, among the mineral manures. I have spoken of the locality at Black Rock, as easy of access, and probably, other places below may be equally so. It was, in connection with rocks of this size, that beds, rich in phosphate of lime were discovered in England, only a few years since, which have been a source of immense profit, to the proprietors or owners.

This formation extends South of Wilmington, more than twenty miles; but, generally, lies concealed beneath a thick covering of sand and clay, and vegetable debris At Rocky Hill, it has been known for many years Nine miles from Wilmington, upon the Railroad, the green sand is twenty feet from the surface. Where the strata of green sand are exposed, in a vertical section, the surface is worn into undulations; and exhibits in consequence of the wearing action of water, which has passed over, rounded ridges, alternating regularly, with depressions. Several in succession occur upon Dr. Togno's plantation three miles North of Wilmington.

As yet, the inferior part of this rock has not been opened in this State. It rests probably upon the granite; but what composes the lower mass is unknown. It may yet turn out, that the mass of pebbles, which border the series of tertiary, upon the west, pass beneath the green sand, and belong to it, rather than to the eocene, where I have placed it.

CONCLUDING REMARKS UPON THE AGRICULTURE OF THE LOWER COUNTIES.

While a bountiful use of manures is one of the main instruments of success, in agriculture, there are others, which cannot be left unemployed The remainder may be comprised, under the head of *tillage*. Proper tillage is necessary, to prepare the way, for the action of manures. When the plow has not broken the turf, or the harrow pulverized and leveled the surface, these fertilizers are left, comparatively, inert. When the land is heavy and impervious, the roots of plants remain unsupplied with nutriment. Where the hoe is neglected, weeds will choke the corn and cotton, and consume the nutriment. Where water stands upon the surface, then, the corn is dwarfed and yellowed by cold. Tillage, then, consists in the use of the plough, the harrow, the roller, and the hoe, and cultivator.—The drain comes in, to complete the work of a perfect system of tillage. That makes the other modes effectual and perfect. I might dwell upon each instrument of tillage, and point out the nature and effect of each operation—but I choose, now, to limit my remarks, to the draining of soils.

DRAINING.

The experience of the most intelligent of the agricultural profession, in this country, and in Europe, is highly favorable to draining. Supported as this agricultural measure is by experience, and sustained by chemical and physical principles, it seems to me, that one word should be said, in recommending the practice, to farmers of North Carolina. I have no reference here, to the swamp lands of the State, or wet lands generally ; there can be but one opinion, as to the necessity of draining such lands, anywhere. But my remarks are intended for those uplands, and meadows, which are always in a condition to be cultivated ; those from which the farmer has taken both corn, hay and cotton, and which do not present any very remarkable necessity of the measure to the eye ; unless it be a preservation of surface water for many days after a rain or showers. Leaving out of view the sandy lands of the Atlantic slope, it will be found that there is, in all the red soils of the State, a very strong affinity for water ;—or, what is better expressed when I say, that *it holds water too long in the Spring.* The effect of this fact is to delay the Spring work ;—to postpone the time when it may be plowed and planted. This is an important matter ; for I think I shall be, and am, in fact, sustained in the position, that a good harvest is generally dependent upon early planting. It may be called in question in the South, where long seasons are the order of the day ; but, when it is considered that all plants obey the seasons, and acquire habits from the impress of the seasons, the rule will hold good for the South as well as the North :—that more may be expected from early, than late planting. The obstacle in the way is the condition of the soil ;—it is wet, and, therefore, cold. It is not so much that the genial warmth of the sun is wanting: but this warmth has to be expended, first, in evaporating a superabundance of water, or in drying the surface. Now, if this water, which keeps the surface cold, is allowed to drain through the soil, and to be carried off by drains, the genial warmth of the sun administers, at once, to the end so much wished, *the immediate warming of the soil.* The principle is obvious and well settled. The *application* which I wish to make, is, to the lands which are regarded as tillable, and capable of producing very good returns; but, after all, they do not give those returns which the labor expended seems to demand.

I design, in these remarks, to call the attention of planters to the question of draining, and then leave the matter for their consideration. I state the object to be, to secure an early planting; though the benefits derived are not confined to that single result. The planter may experiment upon one acre. Drain one acre well; and then compare the condition of the soil of that acre, in April, with an undrained acre adjacent to it.

When I speak of draining, I have no reference to open ditches;—not that they are without their advantages. Open ditches, upon the slopes of many plantations, would be ruinous. I mean, by ditching, a three foot drain laid in tile or stone, and covered. A drain, properly laid, will not wash, or be undermined;—indeed, they will save the surface from gullies, and those dead patches, too common in the South.

COAL FIELDS OF NORTH CAROLINA.

---o---

§ 78. Various opinions have been expressed concerning the Coal Fields of this State. Some, whose opinions are entitled to respect, speak unfavorably of their value and importance. Others, on the contrary, entertain a high opinion of their value. Parties entertaining these discordant opinions have not embraced them without facts which favor their own views, respectively.

Hence, when it became my duty to enter upon an examination, as the Geologist of the State, from whom it was expected that the questions at issue would undergo a careful scrutiny, I determined, therefore, to put those questions to a strict test. It appeared to me that the first point to be established was, that the series of rocks in which coal was known to exist, actually form, of themselves, such a succession of rocks as to entitle them to the rank of the regular coal-bearing deposits, analogous, at least, to what has been determined in other fields, where coal constitutes a part of the series themselves.

To determine this single point, required a detailed examination of all the members of the group,—especially, of those deposited in the immediate proximity of the coal itself. It is, however, never expected that every member of a group will be present in all localities; or that they will, if present, possess the same characters as to color, thickness, mineral composition, &c. Still, certain characteristics will be found in common; and their determination became the first subject for examination.

In order that my readers may be put in the way of forming opinions for themselves, I propose, first, to speak of the common characteristics of a coal field; what rocks belong to a coal field; and what position they occupy; and what fossils they should furnish.

The coal fields of our own country are made up of the following succession of rocks:—

1. Limestone, called the carboniferous limestone;—variable, however, and not always present.
2. Conglomerates;—made up of rounded pebbles, interstratified, it may be, with coarse and fine sands.
3. Sandstone, of various colors and degrees of fineness.
4. Black slates; or slates of various colors.
5. Fire clay and coal.
6. Argillaceous oxyde and carbonate of iron; variable in quantity; generally nodular.
7. Sandstones, and repetitions of the foregoing rocks.

The thickness which the series attain is variable;—in some it exceeds 14,000 feet.

§ 79. The position, in the general series of the coal-bearing rocks, is above what is known as *Devonian*, and extends *upwards* to those known as the *Oolite rocks*, including the latter. They occupy a position not far from the middle of the series of sediments, or *hydroplastic rocks*,—the water-formed or water-moulded rocks. There seemed to have been a period fitted to the production of those plants which form the coal itself.

Observation sustains the view now prevalent among Geologists, that this middle period furnishes the best workable beds ; and that, neither below, nor above, though the sediments are very thick, do they bear coal. This conclusion, I have said, is the result of observation, extended over all the accessible portions of the earth's crust. We do not assign a cause. *It is the fact,* which is important. Still, the fact, itself, implies that the condition of the atmosphere, its temperature and humidity, favored the production of vegetables of those kinds ; and in sufficient abundance to give origin to the coal.

More than half a century has elapsed since attention has been given to the coal-bearing rocks ; and no new country has been visited, but the subject has received attention. The time has passed by when persons who have studied the subject require to be enlightened upon the questions concerning the coal rocks.

What I wish to inculcate, is, that the position which these rocks occupy is not *theoretical ;* but has been determined by observations, extended over a large portion of the earth's crust. The Siberian rocks, for example, are found in America, England, and Russia ; but they furnish no coal.— So it may be said of the Devonian, which is found in these and other countries ; and yet no coal is found in this system.

§ 80. The thickness of all the systems of coal-bearing rocks is great. Room and space are thereby given for containing coal ; and time, also, for the growth of the material which is to be converted into coal. The series in Pennsylvania are several thousand feet thick.

The Geological position of the coal-bearing rocks of this State is another point which required attention. As it is conceded that they are not of the age of the Pennsylvania coal rocks, we have to look about for those which we may satisfy ourselves are of the same age as those in this State, and which have been proved to contain a large supply of coal. We are not left without examples in point, and which establish an important fact, that rocks newer than those referred to are coal-bearing. I allude to those of the Richmond coal field; which have been shown, by Professor William B. Rodgers, to be of the age of the Lias,—of an age newer than the rocks of Pennsylvania.

§ 81. This coal-field, though by no means extensive, when compared with the older, to which allusion has been made, has been productive and profitable; and furnishes seams which are remarkable for their thickness—a seam, for example, which is no less than forty feet thick.

I do not regard this great thickness as a favorable feature; but, as an example of the accumulation of *vegetable matter, in this period*, which proves a geological fact of great consequence,—the adaptedness of the climate of this period to produce the material which has been changed into coal.

We have, in the periods prior to the coal of the Upper Appalachian rocks, negative proof that the climate was not adapted to such results;—for they contain no coal. But, as we approach the carboniferous system, we see signs of preparation; plants appear, which were allied to those which characterised the coal period. These increase, till, finally, they reach the maximum, in point of numbers, in this particular period, when we find all the necessary materials and circumstances for the formation of coal; and it is not until we reach the rocks called carboniferous, that it is found.

Which succeed the latter?—or which can be proved to have been deposited during the subsequent periods?

Again;—the coal-bearing rocks, as they occupy the same relative position, and the same relations as to time, may be supposed to contain a peculiar class of plants, and of animals. All this is true; and it is as much expected to find certain plants and animals, in this series, as to find the sandstones and slates. These facts are well known by miners, who are well informed; and they avail themselves of the facts to guide their examinations.

These facts are so constant, that capitalists do not hesitate to invest money where the characteristics of a coal-field have been determined.

The Appalachian coal-fields, of which the State of Pennsylvania forms a part, are about nine hundred miles long, and about two hundred miles wide; and yet, throughout this great extent of territory, the general characteristics are found the same. And the same may be said of the Illinois and Michigan coal-fields.

§ 82. *Now*, the application of these facts to the coal-fields of North Carolina. The members composing the series, in which coal is known to exist, may be arranged as follows:—

1. Conglomerate, made up of rounded pebbles of quartz, and other hard rocks; united and held together, in the condition of a rock.
2. Sandstones, of different colors, mostly red, of various degrees of fineness.
3. Slate; black, and green, and mottled.
4. Fire-clay and coal.
5. Argillaceous oxyde and carbonate of iron, in nodules.
7. A succession of the series; and finally terminating in heavy beds, and olive-colored sandstones.

In the foregoing series, we find, on comparison, an agreement with those of other coal-fields—the conglomerate, the sandstone, shales, fire-clay, and nodules, of argillaceous-oxyde of iron. All of which, when put together, render it probable that the series actually form a coal-field;—or a series, which are truly carboniferous, or coal-bearing rocks.

The members, then, which may be regarded as common—and, indeed, essential, to a coal bearing series,—are present in the North Carolina formation ;—but, in order to give just confidence in them, I do not deem it necessary to prove that they are of the same age as the Richmond coal-fields. It may be, that the Richmond l eds are not of the age of the lower *Oolite ;* it may be, that they belong to the *Permian*, or *New Bed*.

All the fields furnish one or two fossils in common—a kind of proof which is of some value, as far as it goes.

The question which I attempted to solve, in a way which may be regarded as independent of an actual exploration of the coal-seams, was decided favorably ;—all the facts going to prove a distinct coal series ; though not of the age of the Pennsylvania coal ; but belonging to one during which *coal* has been abundantly produced.

So, when the inquiry is taken up, in detail, though still pursuing a mode independent of the facts accompanying the known seams themselves, we shall come to the same results. The slates, the fire-clay, iron, and the fossils, all point to the existence of coal.

§ 83. Probably more money has been wasted in searching for coal, than any other mineral. It is no uncommon thing for foreign miners, who have just skill enough to take down a breast of ore, to induce the expenditure of capital in sinking shafts, *in any black slate*, without the least regard to the presence of coal The consequence has been, that all the money expended in the operation has been lost. In New York, it has been one of the most common mining failures. Wherever black slate appears, it is perforated, somewhere, with a shaft in search of coal. All these explorations, in that State, have been in the Siberian system, below the coal-bearing rocks. One of the great benefits of the New York Survey, was, the determination that there was no coal there ;

and it has put a stop to the useless expenditure of money in this way. All these mistakes and errors were committed from inattention to the special and general characteristics of a coal-field.

GENERAL OBSERVATIONS ON THE DEEP RIVER COAL FIELD.

§ 84. The Deep River coal field is in the form of a trough! The inferior rocks extend farther than the superior. They may be regarded as beginning in Granville County, in a wedge-form, or pointed mass. The northwest and west outcrop runs, at first, west of south; and passes through a part of Wake, and sends up a short arm to within three miles of Chapel Hill.

The direction of the outcrop has gradually changed to south, 50° west. This direction is very nearly preserved to the South Carolina line. The outcrop is about six miles west of Carthage.

In this coal-field, the uplift has been made upon the northwest side. Its line of demarkation is distinct;—while, upon the southeast side, there is no outcrop. All that is in view, is, the superior rocks, still dipping southwest,—their lower edges being concealed beneath a thick mass of soil.

The dip is slightly variable ;—being, on the south side of Deep River, south 60° east. North of the river, it is south 50° east. At the easterly end, at Farmville, south 10° west. At Hornsville, south 45° west. These last were taken from the coal-slates, where a change has taken place, which is due to the position of the outer and easterly edge of the trough, as it is turning westwardly ; and where the uplifting forces have acted upon the other side, the angle of dip varies from 10°, in the upper strata of the sandstone, to 25° in the inferior beds ; and may, probably, exceed 30°, at some points of the outer edge, near the rocks upon which they repose.

The lithological characters of the whole system furnish considerable variety. But they may be classed as conglomerates ; sandstones, soft and hard, grey, red, and variegated, or mottled ; and green and black slates ; with certain subordinate beds.

The coal seams of Deep River may be described, under three grand divisions, *proceeding* from the inferior, to the superior beds.

1. Inferior conglomerates, and sandstones, below the green and black slates.
2. Black slates, with their subordinate beds and seams.
3. Sandstones, soft and hard, with the freestones, grindstone grits, and superior conglomerates.

§ 85. The lowest and oldest, as appears from the foregoing subdivisions, is a conglomerate. It is formed of rounded quartz pebbles, derived from the neighboring rocks, the gold slates ; and contains the entire series of minerals, which they contain. The most conspicuous part of the conglomerate is quartz. This mineral is rounded by attrition, and occurs in oval masses, rarely spherical, standing out of the rock, in strong relief. These pebbles have, in process of time, become consolidated, without the aid of any cementing substance, and they are so strongly held together, that in breaking the rock, they are broken through, without being broken out, or loosened from their beds.

The origin of these pebbles is evidently in the slates, and from the quartz seams in the slates. This rock being schistose, and largely intermixed with talc and mica, and frequently thoroughly impregnated with pyrites, is subject both to disintegration and decomposition. The quartz by these processes is set free, or disengaged from its matrix — When exposed to the action of waves upon a beach, it is rounded, and while still in their beds are subjected to pressure which results in the formation of this interesting and curious rock.

§ 86. The conglomerates, in their best and perfect forms, are free from soft interstratified matter, which would diminish their solidity, and hence, the mass is exceedingly well adapted for grinding corn, when properly prepared. The beds are two feet, and sometimes three feet thick. The superior are less solid or consolidated; the lower, in their lithological characters, are the perfect millstone grits of geologists. Between the thick bedded masses, there intervenes thinner and less perfect layers, composed of finer materials. These are perishable, and are unsuited to the purposes to which the harder are applied.

The colors are gray, brown, and red; generally, the conglomerates are gray. They have not furnished fossils, except lignite, which sometimes has been found near these lower masses; but even this is never found, except in the softer portions.

The whole thickness of the inferior conglomerates do not probably exceed sixty feet. As a whole, the mass is made up of rounded pubbles in beds of variable thickness, and separated by finer and softer varieties.

This mass rests immediately upon the stratified pyrocrystalline rocks,—the talcose slates, hornblende, gneiss with their subordinate beds, and veins of quartz. They rest upon the edges of the inferior and older rocks—which proves, that the inferior had been elevated more or less, prior to the deposition of this system, and sufficiently to raise up their edges upon which the comglomerates rest.

§ 87. The sandstones succeed the conglomerates, in the ascending order. They consist of variously colored strata, red, gray, and olive. Their texture is not uniform. Extensive beds are made up of the softest of materials, and hence, are constantly undergoing decomposition, and frequently, it is difficult to distinguish the rock from that which has already become soil.

The description of a series of beds will give a better idea of these sandstones, considered as parts of a great formation, than I can convey by any other mode. The following section of the lower sandstone extends North, from Evans' bridge, about three miles, to its junction, with the inferior rocks.

1. The inferior conglomerate is concealed by soil.
2. Sandstone which may be called a hard freestone, dark brown. Near this mass, and to the North, the gold slates appear, which are interstratified with hard green porphyries, alternating with fine talcose slates.
3. Thick bedded brown sandstones, but softer than the preceeding.
4. Gray sandstones or freestones, in which some beds form the grindstone grit.
5. Hard red sandstone.
6. Soft red sandstone, forming a mass, frequently called red marl, but improperly.
7. Gray and olive green sandstones.

§ 88. The series, then, which succeed the conglomerates are made up, as I have already observed, of various strata, some hard, others soft. The predominant color is red, passing into brown. The gray, and those tinged slightly brown, are interstratified with the former; but the former are the most common, and are much thicker than the latter.

§ 89. The slates rest upon the sandstones, I have just described. These slates are thin bedded strata.—They are tender and easily broken, and fall into angular fragments, but not sufficiently hard to form a flat gravel, and hence, from their composition, they are constantly becoming a soil wherever exposed to the weather.

The beds are quite uniform in their mineral characters, and composition. They are green and black; rarely red; but, the latter arises from a kind of bleaching, which they have undergone, by exposure to the weather.

The slate may be described as consisting of

1. Green slates adjoining the sandstones.
2. Black, which frequently alternate with the green.
3. Calcareous beds, in which silica predominates, or in which silica forms more than fifty per cent. These divide the slates into two parts or divisions.
4. Thin beds of impure black limestone.

The coal, fire clay, and argillaceous iron ore, belong to this division of the formation, and the whole series may be described as consisting of

1. Slates of various colors.
2. Coal seams, accompanied with its fire clay, argillaceous iron stone.
3. In proximity to the coal, an impure gray magnesian limestone frequently occcurs, to which succeeds the upper sandstone.

The Fire Clay is a mass of argillaceous matter; quite fine, of a greenish color. Near the surface, it is a clay, and easily moulded, or cut into any form; but at the depth of twenty-five or thirty feet, it is often hard, and more like stone ; still, by exposure to the weather, it becomes soft and shows the character of the element of which it is formed. It is often traversed by roots of ancient vegetables, which gave origin to the coal.

2. IRON.—The iron occur in nodules, or concretions, some of which will weigh between three and four hundred pounds. It is the common argillaceous oxyde, mixed with the carbonate. It does not, therefore, form a regular stratified mass ; but seems to be a constant associate of the coal-seams.

3. THE COAL SEAMS.—There are fine coal seams ; the order and relation of which to the other beds is illustrated by the following section :—

1. The first indications of the coal seam is by the appearance of micaceous sandstone alternating a few times with the slates.
2. Fire clay. Its greatest thickness is about ten feet. A bed of this thickness was perforated by Mr. Campbell, of Moore Co.
3. Coal seam.
4. Slaty beds, with argillaceous iron ore.

It seems to be established by observation, that the fine seams of coal are quite constant.

If we take an illustration from the Farmville and Hornville mines, the following order of deposits will be observed :—

1. Shaly sandstone, with fire clay.
2. Seam of coal, three feet thick.
3. Four feet sandstone.
4. One foot of coal, and fire clay beneath.
5. Shale.

These lower beds dip southwest, and pass beneath the seams of coal at Farmville, which are upon the adjoining lot:

6. Coal, three feet thick.
7. Fifteen feet of slate.
8. Coal seam two feet.
9. Slate ten inches.
10. Coal seam four feet, and fire clay.
11. Slate of various colors with their fossils.

§ 90. There seems to be a slight variation in the materials composing the coal slates. Thus ;

The Taylor mine furnishes the following section :—

1. Slate below the coal seams.
2. Coal seam eighteen to thirty inches.
3. Slate three to five inches.
4. Coal two and a half to three feet.
5. Slate ten inches.
6. Coal four feet.
7. Fire clay, apparently succeeded by gray sandstone and Slate.
8. Soil ; the Taylor mine is three miles west of Farmville.

The eighteen inch coal seam, at the Taylor mine, has been struck ; but not so exposed, as to determine its exact thickness.

The lower seams at Farmville are found at the Taylor mine, which shows a persistence of these thin seams, which are better known at the extreme eastern outcrop.

The seams, at the Gulf, are supposed to correspond very nearly with those already described.

At Willcox's, still farther to the southwest, the seams are not so easily recognized, as the openings are imperfectly made, and were filled with water, when I visited them. The Wilcox seams, it should be observed, are non-bituminous.

The Murchison seams, at the outcrop, contain more slate; but the seam is said to be eight or nine feet thick. The lower seams have not been sought for, at this point.

The fact, that fire clay overlies the highest coal seam, at the Taylor mine, indicates the existence of another seam above it. And, we may expect, that deeper in the basin or trough, others will be struck.

Of the number of coal seams, then, the five noticed in the section, embrace all which have been as yet, brought to light. I shall take occasion, to give my views, hereafter, more fully, of the prospects of this coal field.

§ 91. Slates overlie all the coal seams. Their thickness above them, is about three hundred and fifty feet. The coal strata lie below the midst of the strata beds. The strata above the coal consist of greenish slates, with hard layers occasionally alternating with them. There is, then, taking the slates as a whole, a great sameness in their appearance and composition, and a great degree of uniformity in their thickness, at the different places, where they have been explored. From this fact, it follows, that, they are not to be regarded as subordinate to the sandstone, that they are constituent parts of a great formation, which required an extended period, for their deposition.

I have spoken of the associated coal strata, as slates; but, abroad, a formation possessing their characters would be denominated *marly* or *marl slate*; and the name is highly proper. When their composition and liability to decomposition is taken into the account, they are similar to marls, having lime as a constituent part of their composition

In adopting the term, *marl slates*, we should follow the designation of the English and German authorities; but, as coal is a constant accompaniment of the slates, it seems as well to call them coal slates. It is an important fact, which should be remembered, that the coal is associated with these slates; that, although the sandstones, above and below, are much thicker, and mineralogically, more important, still, they contain no coal; there is no recurrence of slate beds, which can be regarded as repetitions of the one I have described. The shaly sandstones, which are mostly red, sometimes green, but never black, show very conclusively, that, vegetable matter is only sparingly disseminated through the beds, and that the conditions, required for the growth of coal plants, were not repeated, during the era of these sandstones. This fact is practical, and shows in what part of the system we may expect to find coal.

§ 92. I have refrained from speaking of certain unimportant beds, which have been regarded by many as coal seams. I allude to lignite —which is found in the sandstones, just above the conglomerate. It consists, merely, of trees, of the era of the sandstones which have been converted into coal; and, in consequence of the great pressure to which they have been subjected, they have been flattened, and made to assume the form of a thin coal seam.

There is no confidence to be placed in these thin seams—either as seams, or as indications of seams. They are not *coal blossoms;* nor beds of the least economical importance. They are instructing, as geological facts. They furnish us examples of plants growing in an interesting period of the earth's history; and, as historical data, which record the events which have transpired in former times, they are invaluable.

I am thus particular in speaking, in this place, of these unimportant seams of lignite,—or wood changed into coal, still retaining its *structure,*—for the purpose of saying that they are of no value.

There is a still more important point, in this connection, which should be spoken of: It is, that coal occupies certain positions; and I would not allude to it again, in this place, were it not that searches and explorations are being made, at the present time, for *Coal in Lincoln County*,—a field, which, of all others in the State, presents the poorest prospect for finding the substance sought.

The only rocks in which coal will ever be found, in North Carolina, are the sandstones and slate I have been speaking of. The primary slates, though they may be dark colored, and even black, do not derive their color from vegetable matter; but from fine sulphur, diffused through the rock, and which has been derived from the decomposition of sulphuret of iron. Hence, it may happen, that certain black slates may exhibit a feeble combustion upon the fire,—it is no uncommon thing.

The fact should be more generally known in the community, that coal has been formed, exclusively, from vegetable matter; and that coal is found subordinate to certain rocks. Any well informed American Geologist, upon hearing a statement respecting the locality of a supposed coal-bed, can decide the question at once;—at least, so far as this:—the probability of its existence, if the locality is in a district of known coal-bearing rocks; and the *certainty* of its non-existence in parts of country underlaid by rock older than the Devonian system.

Lignite beds occur, also, in the tertiary; and, as it appears, when dried, like coal, it is not singular that it should be regarded as such. My attention has been frequently directed to these beds, on the Cape Fear River, by individuals who regarded them as coal beds of some value.

The outlines of all the geological system are now tolerably well settled, and all that part of the United States, which is underlaid with Primary, Taconic, Siberian or Devonian rocks, are totally destitute of any workable beds of coal. I leave out of view, a debateable ground; that of the upper Devonian, as the late Richard C. Taylor favored the view, that some of the lowest and oldest coal beds may be found in connection with that system.

There are two modes of misleading men, in this matter; the first and most common is by the representation of Cornish and other European miners, who are out of work, or who are totally ignorant of all the essential characters of a coal mine. The second, is from the representations of *clairvoiyants*. To a sensible man, it is scarcely necessary, to guard and caution him, by advice, in regard to the latter. In either case, however, it is proper to ask proof; for it is a matter, which admits of proof,—visible proof; at least, so far as probabilities are concerned. There is evidence, always at hand, for or against. When there is evidence for, it will be proper to investigate its claims, but when there is no evidence, as there can be none in all the rocks older and beneath the true carboniferous, no ignorant miner nor impostor, who pretends to look into a stone, should be allowed to swindle us.

NOTE.—It is difficult to determine the number of seams of coal which exist in this system of rocks. The explorations by boring require to be made, on the line of out crops, before the number can be determined. Five have already been exposed. But as the Wilcox veins are nearly south from Murchison's, and upon one side of the out crop of the Horton, Taylor and Farmville, it indicates that there are seams still above those, and near the junction of the slate with the upper sand stone. The Murchison dips in a direction which will carry it beneath the Wilcox anthracite seams—unless there is an uplift which has brought up the latter from a great depth. At the out crop, the Wilcox is also connected with layers of sand stone and lime stone, which do not elsewhere appear. From these facts it may be reasonably inferred that the coal seams are more numerous than has usually been supposed.

QUALITY OF THE DEEP RIVER COAL.

§ 93. The two varieties of coal, the bituminous and semi-bituminous, passing into anthracite, are known in this coal-field. The bituminous is scarcely equalled for fineness and excellency, in this country, and it has been said by a gentleman, who is well acquainted with Liverpool coal, that it will burn twice as long. A direct comparison has not been made, to my knowledge, but that the assertion has much truth in it, I have no doubt

The Deep River coal is, in the first place, quite free from smut; it does not soil the fingers, but in a trifling degree. It burns freely, and forms a cake; or it undergoes a semi-fusion, and agglutinates, and forms a partially impervions hollow cake, within which combustion goes on for a long time. When a small pile of it is made upon the ground, it may be ignited by a match and a few dry leaves or sticks. It may be ignited in the blaze of a lamp or candle. The coal is, therefore, highly combustible, easily ignited and burns with a bright flame like lightwood, for a long time. It may be burnt upon wood fire. It may be burnt in the common fire-place, and it is not a little strange, that gentlemen, who have used it for many years, in a black smith's forge, should not have used it in their parlors, in-stead of green black oak.

This coal is adapted to all the purposes, for which the bituminous coals are specially employed. Thus, for the manufacture of the carburetted hydrogen, for lighting streets and houses, there is no coal superior to it. It will require less expense for furr ishing it; because, it contains so little sulphur, from which sulphuretted hydrogen is formed So, also, in the grate, it will be far less offensive, for the same reason. But, as it is rich in bitumen, it will furnish a large amount of gas, and that which is, comparatively, pure. This advantage is one of great importance. It should, also, be stated, that it furnishes an excellent cake, which may be used for manufacturing purposes, and as it is left very porous, it is in a condition to absorb a large quantity of the solution of cyanide of potassium; and hence, is well adapted to the work of reducing the metals. It is scarcely necessary to add, that it is admirably adapted to steamings, inasmuch as its flame is free and durable. For forge use, it is not surpassed by any coal in market; and for parlor grates, it is both pleasant, economical, and free from dirt. If a chimney has a poor draft, it is liable to the objection common to all coals of this kind,—the escape of soot into the room.

The qualities of the Deep River coal are of that character, then, which will give it the highest place in the market. The localities which have been best explored, and where coal of a decided character has been attained, are at Hornesville and Farmville, both in the same neighborhood. The Taylor mine, the Gulf or Horton, and the Murchison mines, all furnish a bituminous coal, which may vary in some minor points, yet is quite similar as a whole. The Horton mine has been used the longest. It was known in the revolution, and a report made to Congress, respecting it, is still extant. Had the propositions or views been carried out, which were expressed in that report, we can scarcely tell, what the results would have been, not only upon the population of Deep River, but also, upon the enterprize of the State. It must be noticed, that Deep River is central, and in the interior of a country, abounding in iron; that it is navigable, by aid of certain improvements; that it commmunicates with the ocean, and finds a market abroad, for a surplus of the products of manufactures and agriculture; that a use of the natural advantages, to a partial extent only, makes a home market.

But the time had not come, for improving the resources of this district. They are, therefore, reserved entire for the present, and they cannot be neglected longer, unless a suicidal State policy is pursued.

But however fine and excellent a coal may be, it is necessary that it should form extensive beds, in order to have a commercial value.

§ 94 The next question, then, of interest to the community is, (for the community is interested as much as the owners,) will it bear mining. and the expenditure of the necessary capital, to take it to market. To answer this question, it is necessary to make some calculations, by which we may form some just view of its quantity ? In doing this, we may venture to assume, on a geological basis, that the coal seams, which outcrop from beneath the sandstones, xtend beneath them, and for what appears to the contrary, the slates, with their coal becs, are coextensive with the under and overlying sandstones. his formation is known to form a belt of rocks, from 12 to 14 miles wide. The line of outcrops of the slates, upon which coal has been raised, is about 20 miles But the line of outcrop of the unexplored slate, which embraces the coal, is at least 60 miles within the State, on a line running south of west. We may assume the following data, viz : that the coal beds extend from their northern outcrop, three miles beneath the sandstone ; which is about one third their natural extent ; and that the line of outcrop, upon which coal is, and will be found, is thirty miles. If the thickest seam of coal is worked, which has a thickness of 6 feet, exclusive of a thin band of slate, it will give for every square yard of surface, two square yards of coal. A square acre has 4.900 superficial yards; hence, there will be 9,800 square yards of coal, in each acre, and as a square yard of coal weighs a ton, there will be for every acre, 9,800 tons of coal. A thousand acres will give 9,800,000 fons of coal, or a square mile, 6,272,000 tons. This coal field is known to extend thirty miles, in the direction of outcrop, and to be workable, for a breadth of three miles. We may from this data, calculate how much accessible coal we may expect to find, in this quite limited field. If the field covers only 43 square miles, the lowest estimate to be taken, we may calculate its value, by the following mode:

If one hundred tons of coal are taken out daily, thirty thousand tons would be removed annually, reckoning three hundred working days to the year. It would, at this rate, require over three hundred years, to remove the coal from a thousand acres, or, over two hundred years, to remove that which underlies a square mile, or, eight thousand six hundred years, to remove the coal of forty three square miles. If in estimating the value of this coal field, we base our calculations upon time, they should satisfy us; or if we base them upon quantity, they will warrant the investment of capital. In these calculations, we have both time and quantity, and the State, in encouraging improvements as well as individuals, may look forward with confidence, on the permanency and safety, in investments, in this kind of property. The wants of the world are with the population—indeed, they keep ahead of simple increase of individuals. The quantity to be removed annually may be increased, and leave the time sufficiently long, to satisfy the investment of capital; or the time may be increased, by diminishing the quantity, and still the annual profits of the investment should satisfy the capitalist. But while population increases at a rapid rate, the resources of the forest for fuel are diminishing at a greater ratio, than the simple increase of population; therefore, there is no way in which capital can be so safely invested, as in coal lands.

If the foregoing calculations are correct, they justify the work which has been undertaken to improve the navigation of Deep River. It is prudence, to be cautious in schemes of this kind, but in this case, the amount of property beneath the surface or in the rocks, upon this river, is enormous—it should be dug out; and what it costs to do this, will be turning materials and labor into money. If the whole enterprize is begun, and carried on in a proper spirit, every nook and corner of the State, from Currituck to Buncombe, will feel an invigorating influence.

But the calculation, as to the quantity of coal, will probably far exceed, than fall short of the estimates. In the first place only a part of the area is taken into the calculation, and then, in assuming the thickness of the principal beds, as only six feet, it may be regarded as only the minimum thickness. It will rather increase than diminish; this view of the matter is supported by observation. For as the slopes have been carried along the dip, there has been a perceptible increase already. It is also to be considered, that at the outcrop, when vegetable matter forming the coal is only upon the outer vein, it should be twice that at a distance from the outcrop; for we may suppose, that, in the middle only, of a coal basin, do we obtain the maximum thickness. Thus, one of the coal seams in the Richmond basin is forty feet thick. The Deep River beds, not having been broken up, or not having suffered an uplift through the middle of its trough or basin, exhibits no where near the surface, an outcrop of coal, except *upon the rim*, or outer edge of the basin. As we penetrate into it, we have grounds, which justify the view, that the seams will increase steadily in thickness, as the slope penetrates into the basin, towards the centre, and then the seams, which now appear only upon the outer rim, will thicken, and perhaps unite and form one distinct heavy seam towards the middle of the basin or trough.

The foregoing views as to quantity are founded upon data derived from observation, the phenomena of coal fields, and theory, which is well sustained, respecting the manner in which successive seams of coal have been formed.

The calculations as to the quantity of coal in the Deep Ri- coal field are based upon what is known, and without reference to what we may possibly find by exploration hereafter. These calculations must be regarded as satisfactory, and such as will justify the hopes and expectations of the owners, and those who are interested in the improvements of the day.

———o———

THE SANDSTONES ABOVE THE SLATES.

The upper division of this system of rocks is made up of a series of sandstones, which differ only slightly from those below. They are hard and red, brown and mottled, or variegated rocks, which are frequently separated from each other by soft greenish clays. There are to be repetitions of the inferior members, so far as this composition and origin is concerned. Among the different red and brown strata, it is very common to find those which contain many small irregular cavities. Probably these cavities contained imperfect crystals of salt, or other soluble substance, which has been removed by solution.

These sandstones, like the inferior, are destitute of fossils, excepting a few obscure vegetable casts, which are indeterminable. Near and just above the coal slates, a stratum of impure limestone is filled with the posidonia minuta which disappears with the commencement of the red rocks.

Among the softer layers oxyde of iron occurs in small nodules. It is not in sufficient quantities to be of much account. It has been employed in dying woollens, and hence has received the name *of dye stone.*

But, so much of the upper sandstones are concealed by soil, that even plaster might be hid within the strata.

———o———

AGRICULTURAL CHARACTERS OF THE UPPER AND LOWER SANDSTONES.

§ 95. The softer strata disintegrate, and pass into the condition of soil. These soils are always red, but the red inclines more to a brown than the soils, which are formed from the gold slates, which, from the presence of sulphuret of iron, become red also. The sandstone soil is derived mainly from the argillaceous parts of the formation, and hence it bears the characters of an argillaceous soil, a degree of stiffness which suits the cotton and wheat plant. Its stiffness and comparative impermeability give it retentiveness: but it also holds too much water, and especially where the land is flat, it is far too compact to admit of high cultivation without the aid of draining. The elements of the soil are adapted to the highest and best cultivation. But as the surface water is compelled to flow over the surface, and into channels of its own forming, it is very liable in the end to score the soil deeply with gorges, and ravines, and thereby injure very much the plantation. These red soils, also, when cultivated to exhaustion, or even before they become bare, render it extremely difficult to restore to the surface its lost vegetation. The only mode in which it can be effected, is to cover and protect those places with straw or brush. The heat of the sun should be softened, and the surface protected from washing anew by rains. By persevering in this way, these ill looking and barren spots may be removed from a plantation.

THICKNESS OF THE FORMATION.

§ 96 It is difficult to obtain the data by which the exact thickness of this formation can be determined. It is concealed by its own debris so generally, that it is only from the subordinate parts that its thickness can be inferred.—Taking the subordinate beds, dip and width all into the account, the whole thickness of the formation cannot be less than five thousand feet. The inferior mass, or that below the slate, is about fifteen hundred, the slate five hundred, and perhaps six hundred, and the superior division three thousand. This last estimate is below the mark rather than above it. In forming an estimate of the thickness of this formation, I have been careful to guard against deception which often arises from a repetition of strata in consequence of a succession of uplifts. In this formation the danger may be greater than in many others, as the strata are concealed by soil. But the indications upon the surface go to prove that the strata have been disturbed only at one period. A single trap dyke traverses the entire formation from the South-west towards the North-east. This trap appears at numerous points along the line, as at the Gulf, and Evans' mill, crossing the river twice, and forms at each place a considerable ledge of rocks The falls are due, therefore, to the trap dyke. At another place near Evans' mills, in the road by Mrs. Roberts,' the dip of the strata is reversed for a short distance. But this is only local, and does not affect the great mass of strata of which the formation is composed. South of the river, they preserve a great uniformity of dip, as to direction, and amount, and no part has come to light, which indicates an instance of repetition.

§ 97. Having briefly noticed and described the three members, composing this formation, a few remarks upon the origin of the materials will be in place.

The materials, considered in their mineralogical characters, were derived from two distinct sources: the common source was undoubtedly the slate of the gold series. Of the correctness of this view, there can be no doubt, as they even contain a small quantity of gold, and the quartz retains the peculiar characteristic of that of the slates; it is best seen in the conglomerates, where the masses of quartz are larger than in sandstone. The color of the sandstones is due to the presence of oxyde of iron. The iron came from the same source as the sandstones. These slates are highly charged with pyrites, which have been thoroughly decomposed;—the sulphur dissipated, or probably it has formed new combinations. The oxyde of iron, being insoluble, has remained with the particles composing the rocks; and it has formed, of itself, no combination with other bodies, unless it is with carbonic acid. The iron forms a superficial coating upon the angular grains of sands, and may be removed by washing, when it leaves a perfectly pure white sand. In the other case, it is intermixed with clay and a smaller proportion of fine sand, and forms the softer beds, which are sometimes called *red marls.* It seems, therefore, that the materials of this group of rocks were brought from the northwest side of it, or from that portion of the gold rocks which lie in that direction. As, however, the area of deposit must have been basin shaped, or trough shaped, the materials must have been derived, also, from all sides of it; but our examinations are facilitated upon the northwest side by a vertical movement; hence, we are able to test the truth of these views, rather upon that side, than upon the southeast.

§ 98. It is more difficult to determine the origin of the materials composing the coal slates. These contain lime, but no sand, except in combination with alumina. The slates, therefore, differ so much from the sandstones, that there can be no doubt, that the origin and source, from whence they were derived, was different also. They indicate a total change in the direction in which the sediment came, and as there is no formation to the northwest, or southwest, to which their origin can be attributed, I am inclined to regard the source as concealed by the present ocean.

The sandstones in the United States, of the ages closely approximating to those under consideration, were deposited in shallow water. The ripple marks, the foot prints, which have been preserved, are regarded as proofs of this position. Taking the same ground and kind of evidence, I have not yet been able to furnish the same ground of proof for their deposition in shallow water. I have not yet discovered footprints, or ripple marks, and the only fact, which goes to support the view entertained, of the depth of water, in which the strata were deposited, is, a single layer, which had cracked in drying. These cracks were subsequently filled with sediment. Still, but few opportunities for a disclosure of the interior of these deposits have been furnished. The blocks of sandstone, which have been employed in building, have been procured from the surface, and which were already loosened from their beds; besides this, the strata have not been laid bare by an inundation. The means, therefore, for obtaining a series of facts, bearing upon these questions, are insufficient. They throw but little light upon them, and we must, therefore, wait until railroads and internal improvements have laid open the series to inspection.

The evidence, as it now stands, favors the view, that these rocks were deposited in deep water; the absence of all ripple marks, the great thickness of the deposits, and thickness of the deeper layers, indicate for the sandstones deep oceanic deposits.

AGE AND EQUIVALENCY OF THE FORMATION.

§ 99. The age of a rock, or of a formation, is determined without difficulty, provided it contains fossils; especially those belonging to the molusca;—or, if its relations to other systems can be seen; whether those systems are above or below. But the sandstones and coal-stones rest upon a series of gold-bearing rocks, whose age dates back farther than any fossiliferous rocks;—hence they cannot be employed for comparison. The only rocks which rest upon them are the sands of the tertiary. We have no way boards by which their relations to other formations can be determined, excepting those of the most general character.

The fossils are exceedingly scarce; and their species few in number, and not very distinctive. They are confined to one species of molusca: a small *posidonia* or *cypris;* which is regarded as a *crustacean,* and which is only the size of a grass-seed; the teeth of two or three saurians, and the scales and teeth of one or two fish. The posidonia does not differ from the posidonia of the Richmond beds, except in size:—it is smaller, and resembles the *P. minutiæ* of Goldfuss.

This fossil is remarkable for its numbers; every layer in portions of the slate is crowded with them, and they range from the top to the bottom. It is usually one-eighth of an inch in diameter; the largest rarely ever exceed one-quarter. It is always flat, in the slates, from pressure, and always round and plump in the thin beds of impure limestone. The cypris, or it may be a cytherina, is about one line in length and pointed at both ends, and smooth. It resembles a hay seed; and so numerous is this minute fossil, that thick layers are made up almost entirely of it. It is scarcely possible to touch a point with a pin, and not mutilate one.

The presence of the cypris indicates that the slates are a fresh water formation—they go to prove that a remarkable change took place after the lower sandstones were deposited—that the ocean, in which the sandstones were formed, was removed—and the basin, to which sediment had been brought from a distant quarter, ceased to be brought to it as formerly, and from the same as they had been;—and, in fine, that what had been sea became a fresh water lake. It was not until this change from red sandy sediments took place, that the fossils peopling the waters appeared; and then they were confined to a very limited number of species

If, however, this small fossil is a cytherina, the change which it is supposed may have occurred at the close of the oceanic sandstones, was only in the direction of the sediments; a change which appears to have been sudden—for the sandstones scarcely alternate with the lower slates. They begin, as it were, at once; but the basin or trough was still oceanic.

Of saurian remains, in the formation, I can speak only of two species in the sandstones:—one below the slates, and of the crocodilian type; and one above, with long curved teeth. And, probably, three species in the slate—the teeth of one are long, slender, and curve four inches; the teeth of another, of a medium length, and only slightly flattened, and very finely serrated on one edge; the teeth of the other, distinctly serrated on both edges, and agrees with figures of the *thecodonto saurus* of Owen.

The latter points to the *Permian* age; and Sir Charles Lyell has observed, that this *saurian* was regarded as the oldest animal known of that type;—and, from its presence in the older deposits, Mr. Owen has shown that it militates against the doctrines of the Author of the Vestiges of Creation. It ranks with the highest animals of that type; proving that rank is not determined by the periods in which animals have lived. The most remarkable saurian, if saurian it is, is the species furnished with the long slender teeth, of which I have seen no figures resembling them in form or length. They are slightly flattened, giving an oval in a transverse section; but their sides are not armed with serratures.

The bones in the rocks above the coal, are black. The sandstone had become concretionary, and exfoliated in their coats, like those of an onion; and hence, it was impossible, to obtain them in a good condition; besides, the rock had, also, become exceeding hard and tough.

The fossils, being in my opinion, new, throw no positive light upon the age of the rocks, in which they occur.

The fish scales are quite small and smooth. Their form is rhomboidal, some acute, others obtuse. The teeth are small, slender and pointed, and seem to fork slightly, at their roots. Another fossil, which might be mistaken for a vegetable, is, undoubtedly, an appendage to a fish.

The vegetables are few in number, and differ from those of the coal rocks of Pennsylvania, or the flora of the carboniferous system. An Equiselites, differing from E. Communis, is the only one of this genus I have seen. A lycopodites, and other allied forms, are all I have yet found, except a naked aud rather spinous vegetable, which is unknown in the carboniferous rocks. It is a cellular crytogamous plant. This is very common and abundant at Madison, and one or two layers of slate are covered with it at Evans' Mills.

The roots of vegetables, in the fire clay, are thin, narrow, ribbon-like tissues; and have lost their vegetable structures. Their thinness and compressibility show, however, that the roots were spongy, of a loose texture, and were aquatic.

The meagre list of plants and animals, then, deposited in the slates, furnish only grounds for conjecture, to what age the formation belongs. My opinion, derived from all the facts and circumstances known to me, inclines me to adopt the belief, that it is the upper new red sandsstone Still, if the Richmond coal basin is of the same age, as the coal rocks of North Carolina, Geologists will be disposed to place the series along with the Oolites or Lias, a [illegible] Wm B. Rodgers and Sir Ch. Lyell have done—on the ground, that the fossils are, in part, identical in species with those of Whitby, in England, where those rocks are well developed. Mr Lyell observes, that the sandstones containing fish, of the Connecticut river, are of an older date than the strata containing coal near Richmond. The higher antiquity of the Connecticut beds cannot be proved by direct supposition ; but the fact is presumed from the structure of the country. That structure proves them to be newer than the movements to which the Alleghany chain owes its movements or flexures; and this chain includes the ancient coal formations among its contorted rocks, The unconformable position of this new red with the primary is often seen. He regards the sandstones of the Connecticut valley as *triassic ;* but, to what portion of the *triassic*;, which division, whether upper or lower, is not determined.

In Europe, the *triassic* is rich in fossils ; and different parts of the series are so well characterised by fossils, that the determinations are not difficult. But here, in this country, the Connecticut valley, the New Jersey beds, the sandstones of the Potomac, and Fredericksburg, and North Carolina, are all obscure, from their relations, and from the absence of characteristic fossils.

THE DAN RIVER COAL FIELD.

§ 100. In Rockingham and Stokes Counties, a series of rocks have been known for a quarter of a century, as coal-bearing. These rocks are similar to those of Deep River, and consist of the same members. They lie in the same order, and have the same relations to each other, as those of Chatham and Moore, or Deep River.

While there can be no doubt respecting the age and relations of the entire series, compared with those of Deep River, still I have observed a few peculiarities worthy of notice.

The Dan river coal deposits may be divided, for the convenience of description, into five parts:—

1. Imperfect conglomerates and breccias.
2. Lower sandstones, including the soft and hard.
3. Coal slates, with their subordinate deposits.
4. Upper sandstones; including the soft and hard kinds.
5. Conglomerates; or brecciated conglomerates.

The several parts, constituting a complete and perfect system, occupy a synclinal trough, and lie in the primary or stratified pyro-crystalline rocks. Its direction is north-east and southwest. The axis may be defined by uniting Leaksville and Germanton by a line. This line will represent the direction of the coal slates.

The general dip of the system is to the northwest;—the angle of dip lies within 15 and 40°. The dip is usually above 20°. In North Carolina, the rocks extend 40 miles. The breadth is between four and seven miles. The system extends into Virginia on the north; but how far, I am uninformed.

This field, it will be observed, covers a smaller area than Deep River. It is similar, in some respects, to the Richmond coal fields, but is disconnected by the intervening primary rocks.

If we consult a map of the United States, and mark upon the map the position of the coal-field to which reference has been made, we cannot fail to notice the singular fact, that there are three small troughs, formed in synclinal dips of the primary slates, and all lying with their axes directed to the southwest, or nearly parallel to the present Atlantic coast.

These troughs are now disconnected, and an examination of the series, their outcrops, &c. go to show that each was formed in a trough by itself, and totally disconnected with each other. Each was formed in the bosom of its own sea, and each remarkably deep. The areas upon which these rocks were deposited, have never suffered from denudation, or from great fracture; but are traversed by moderately sized trap dykes. The Richmond coal beds have been disturbed more than those of Dan River, and the Dan River lie inclined at a greater angle than Deep River.

When our examination is extended to the Hudson and Connecticut Rivers, similar rocks are found. The sandstone is accompanied with conglomerates and slates. The latter, however, are hard, and retain the impression of the fish and fossils better than those of the Dan or Deep River. All these beds of sandstone lie parallel to each other. They are comparatively long, but the breadth is inconsiderable. That these several isolated series represent one period, is highly probable, though not geologically proved.

Four of these isolated troughs are characterised by outbursts of the pyroplastic rocks, or igneous. The Trap or Palisades of the Hudson ; the vast fields and mountains of trap in the Connecticut Valley, extending more than a hundred and fifty miles ; and the heavy trap dykes of Deep River, and the minor trap dykes of the Dan, belong to one era. They all cut through the sandstone and slates, and send lateral branches of the once molten mass both between and upon the layers, baking and hardening those which are in contact or in proximity with them.

Geologists are now very much inclined to adopt the view that outbursts of igneous matter, though at distant points, but found upon and through the same formation, happen at one and the same period.

Proceeding still farther North, our attention will be arrested again by a still more extensive outburst of trap in Nova Scotia. It is not satisfactorily determined whether the traps of Nova Scotia are connected with the new red sandstones. Still, it seems to have happened at a period subsequent to the carboniferous ; and the trap lies upon, and has intruded itself into, a rock, whose mineral characters are similar to those of New Jersey, Deep River and Connecticut.

But the foregoing remarks may be regarded as digressions. My object in these remarks is, to *identify age by means of phenomena*, and show that, when rocks possess characters in common, and where certain phenomena are of the same kind, and are observed to be common to them also, it is an indication that the rocks belong to the same period.

It is remarkable, too, that all these troughs of red sandstone repose directly upon the primary rocks. The junction of sandstone with primary is very distinct at Blomidon, Nova Scotia, in the Connecticut Valley, in the Hudson River Valley, in the Richmond basin, the Dan River, and the Deep River. The most Southern troughs of red sandstone are the least disturbed, and the smallest quantity of trap has been ejected. In Nova Scotia it has reached its maximum. The whole outburst has extended through twenty degrees of latitude.

Whether the foregoing facts do really prove that the systems are one in age, and belong to the same period or not, may still require proof. The facts themselves are interesting. We may require many additional particulars to enable us to interpret phenomena aright, and assign to those sandstones and slates their true age.

§ 101. The lowest mass upon the Dan River, which belongs to the sandstone series, is a conglomerate, quite imperfect, at least, where it has fallen under my observation. At Leaksville, I have not seen the lower conglomerates ; but at Germanton, an imperfect mass, occupying the lowest place in the series, is formed of angular pins of granite, mixed with a gray and reddish sediment, in very unequal proportions. Its appearance might easily deceive an inattentive observer. It has an exceeding close resemblance to some varieties of granite. After a close inspection of many large rocks lying near the small creek at this place, rounded pebbles were found ; and, by still farther search, roots of trees in beds of lignite were found, also branching into the rock.—This represents the fine beds of millstone on the Deep River. The mass is thin, and of little importance.

Immediately above this bed of brecciated conglomerate, there is one of the finest exhibitions of an ancient forest in this country. It consists partly of roots of trees changed into lignite, and partly of perfectly silicified trunks of trees, exceeding two feet in diameter. The soil in which the majority of these trees grew, is still concealed. Segments of their trunks stand out of the soft rock, inclining at an angle to the horizon, but lean in a direction contrary to the dip of the rock. A road cuts through the strata in which the forest grew. All that remains of it are the trunks ; it was impossible to find a leaf or stem of herbage or fruit. The softer and more perishable parts and organs are destroyed by unknown agencies. Perhaps some fortunate blow of the hammer may bring to light the leaves and fruit. The structure of these trunks prove them to belong to the natural family of coniferæ, or the family to which the pines, spruces and hemlocks belong

The trees extend for half a mile or more, and no one, on seeing the number, can doubt that here grew a forest when the rocks were forming.—Similar trunks have been found at Madison, and pieces of trunks occur upon Deep River, near Evans' bridge, and another forest of the same character upon Drowning Creek, in Richmond County. They occupy the same position in the series. These trunks are geologically important, and may be employed to assist in identifying the system with any other at a distance. Numerous fragments of trunks, also, occur in all the subsequent formations, especially with tertiaries, and in the superficial cutting for rail-ways. I was at a loss to account for their occurrence in positions where agencies could not be supposed to exist, competent to silicify wood. I have been satisfied that most of the scattered trunks were derived from the red sandstone formation. They have been transported by rivers, and by various agencies, which have also carried the slate rocks, and deposited them in the green sand, and the various subsequent beds of the tertiary. Their direction of transport is eastwardly.

§ 102. In connection with the strata I have described, (the brecciated conglomerate,) there occurs no clay or argillaceous formation, which has a perfect concretionary structure. Large concentric circles are formed ; some of which are two feet in diameter. This part of the rock is extremely soft, and is nothing more, nor less, than clay of a light green color. It is rare to find a series of perfect concentric circles, and terminating in a nucleus of the size of a two shilling piece, as at this place. They are due to molicular movements, which have taken place, subsequent to the time of deposition. We are obliged, from phenomena of this kind, to reckon molicular force, as one of the silent geological forces, which have been instrumental in effecting important changes in the earth's crust.

§ 103. These argillaceous beds lie beneath the common sandstones of the formation, which consist of variegated and gray masses of rock. They terminate with the coal shales. At Leaksville, a hard silicious slate intervenes between the lower beds of sandstone and the slates. It is bluish and flinty, approaching, however, a sandstone in its composition. It is at least two hundred feet thick. It also contains a few layers, which externally resemble a mixture of carbonate oxyde of iron. It was in one of these layers, I discovered the fragments of the skeleton of a saurian.

The middle part of this formation of sandstone is occupied with a soft marly slate—the coal slate of the system. It differs in no respect, from that of Deep River, bearing the same fossils, the posidonia and cypris, in equal abundance, through all the strata, of which it is composed.

The coal beds of Leaksville lie in these slates; the beds in which the coal seams are exposed are two miles from the village, on the plantation of Mr. Wade.

The coal appears in a long ridge, rising about sixty feet above the meadow, which lies in the bend of the Dan, at this place. The following section is partially exposed at Wade's coal mine —

1. Shale below the coal seams.
2. Shaly micaceous sandstones, two feet.
3. Shaly coal at the outcrop, eighteen inches to two feet.
4. Micaceous shale, two feet.
5. Coal, two to three feet.
5. Shale, 110 feet.
7. Seams of a hard blue magnesian limestone, intermixed with silex, four to six feet.
8. Soft, green, bluish and black shales, filled with posidonia, sixty feet.

The shales still continue covered with soil; the thickness of the shales is not less than five hundred feet.

The hard calcareous layers are separated by slate at the surface.

The calcareous layers lie above the coal seams, and as they extend nearly if not entirely through the formation, they may be observed as way boards in searching for coal. The same layers appear in Madison, and contain abundance of septaria of the size of a goose egg.

Dip of the co l slates at the coal mine: N. 35° W.; angle of dip 25°; strike S. 80° W. It is not improbable, that the angle of dip will diminish as the seam is penetrated.

§ 104. The section of rocks lying between Eagle Bridge and Gov. Morehead's factory, is exhibited in the following section; though, it should be observed, that the rocks are concealed, at some points, between the bridge and factory.

1. Sandstones and conglomerates, concealed at the bridge.
2. Flinty black slates, two hundred feet, with saurian remains.
3. Coal slates, consisting of the usual green and black slates, with the posidonia and cypris, and a few obscure species of plants, (*Lycopodiaceae*,) fifty to six hundred feet.
4. Red and gray sandstones.
5. Conglomerates.
6. Shaly and green variegated sandstone.
7. Conglomerates, at least five hundred feet.

These conglomerates are hard, and contain many angular fragments, or those which are but slightly rounded; and some of these fragments are quite similar to the flinty slate below. The beds resemble hard gray wackes of New York, except that the masses of rounded quartz are much larger. The superior beds of sandstone occur at the factory, and have been employed as a building stone.

§ 106. The series of sandstones which lie between the bridge and the conglomerates, are better exposed upon Factory creek, about four miles from Madison, on the road to

Martins' lime kilns. The latter predominates at this locality. The creek has uncovered the strata, for half a mile.

The section upon Factory creek is represented by the following strata, the strike of which is S. 70° W. dip 20°.

1 Soft greenish slates and shales above the coal slates.
2. Coarse sandstone with pebbles.
3. Red and brown sandstones.
4. Porous red sandstones, or sandstone with angular cavities, similar to those in other rocks, which have contained a soluble salt.
5. Green and gray hard sandstones.
6. Coarse sandstones containing rounded pebbles.
7. Conglomerates similar to those at Gov. Morehead's factory, at Leaksville.
8 Soft sandstones, like the red marls.
9. Slates with quartz veins, dipping beneath the sandstones.

The thickness of the series above the coal slates, is between four and five thousand feet.

At Madison, the series below the coal slates, on the East side of the Dan, at the new bridge, is represented by the following sections :

1. Gneiss dipping beneath the sandstones.
2. Soft variegated sandstones, with mica, and imperfectly bedded ; at least two hundred feet thick, east of the site of the bridge.
3. About one thousand feet of green shaly sandstone, with drab colored sandstone, interlaminated with the series ; strike N. 65° E. angle of dip 45°.
4. Red sandstones, with cavities.
5. Green and dark colored coal shales.

§ 106. The coal has been exposed about four i ches from Germanton, on the plantation of Mr. Mathews. The strata as exposed, are arranged in the following oider.

1. Slate below.
2. Fire clay.
3. Coal, eighteen inches.
4. Slate, one foot.
5. Coal, eighteen inches.
6. Black slate, five feet.
7. Sandstone and state.

The coal at the outcrop is not pure, or it contains some pyrites. Still, at a new locality on this plantation, discovered by Dr. McClenahan, at the time of our visit, the prospects are better than at the shaft, where the coal was first taken out.

The attention, which has been given to the Dan River coal field, has as yet been too inconsiderable, to develope its riches. It appears, that fro.n Leaksville to Germanton, coal s exposed at several points, besides at the extremes of the formation, leaving out of view its extension into Virginia.

§ 107. The foregoing descriptions of several subordinate sections will convey to the reader a correct idea, (so far as description will convey.) of the conglomerates, sandstones, and slates of the Dan River coal field. From the observations which I have made, I am inclined to regard the conglomerate as the least constant mass, and the most variable in its characters. It exists at Germanton, but is imperfectly developed ; whilst at Madison, it is replaced by a soft mass of the red sandstone.

At Leaksville, and also, not far from Madison, this series contains some remarkable beds of brecciated conglomerates which are probably absent or wanting upon the Deep River.

The shales or marls, appear to be the most constant mass. It preserves its thickness and all its characters unchanged, and most, if not all the subordinate beds, are developed, both upon the Dan and Deep Rivers.

The fossils invariably appear wherever the slates are found ; so, also, the impure limestones, with their concretions, are equally constant. though they are quite inconsiderable in mass. In the Dan River coal field, the lower rock if fully disclosed is much thinner, and less important, than the same mass in the Deep River, in which, as I have already observed, the slates seem to be equal in importance in each. The conglomerates of Deep River are very prominent, and quite important.

But, if we compare the thickness of the sandstone above the slates, they seem to be thicker, and more fully developed upon the Dan. I am also inclined to estimate the entire thickness of the sandstone series, as greater on the Dan, than upon the Deep River.

PRODUCTS OF THE UPPPER NEW RED SYSTEM, OR TRIAS, OF THE DAN AND DEEP RIVERS.

§ 108. 1. Fire Clay.—The fire clays of the Trias are well adapted to the manufacture of fire bricks. The clays connected with coal seams have long been used for this purpose ; and hence the name, *fire clay*. The clays being free from iron, lime, and magnesia, are highly refractory in the fire; and hence are well adapted, from their composition, for the manufacture of such articles as are required or designed to be subjected to a high heat. The seams of fire clay are, in a few instances, ten feet thick ;—others are only two feet. They are always found in seams, subordinate to the slate. Some seams of fire clay do not bear coal at their outcrop. The material is very fine and even-grained ; the silex is never coarse or concretionary. The only obstacle which stands in the way of mining this clay, is, that it often becomes hard in the deeper parts of the seam. It is abundant upon the Dan and Deep Rivers. It is entirely distinct from the ordinary clays of the country. It is confined to this formation; and is, in fact, a subordinate part of it: and is never absent, it is said, wherever a coal seam exists:—though it may occur, and coal be absent. It appears to be fine enough for the manufacture of articles much finer than fire brick.

2. Oxyde of Iron, or Argillaceous Carbonate of Iron.—The coal series appear to furnish always more or less of this variety of iron ore. It occurs, usually, in nodules, from the size of an egg up to a barrel. Generally, their form is an oval or flattened sphere. The strata are parts of the coal series, and subordinate to the formation, and are depended upon, to a great extent, for the supply of ore for iron. Its qualities, especially when manufactured with coal from the beds, is not of the first order; but, as it is made into iron cheaply, it is valuable ore.

3. Limestone.—The limestone, which has hitherto been exposed in mining, is of an inferior quality, and only small in quantity. The layers do not exceed a foot in thickness. At the Wilcox Mine, and at an opening on the plantation of Mr. Campbell, on Deep River, layers of tolerably pure gray and granular limestone occur.

The seams and thin beds of limestone lying in and dividing the slate, is impure from silex, and is probably magnesian. Septaria are formed in this band, which, taking the whole, and including some intervening slate, is from four to five feet thick. It may serve a good purpose in making hydraulic lime. It should be tried. It occurs at Leaksville and Madison. The Deep River band, though it occupies apparently the same position, seems to be more silicious than at the other places mentioned. When limestone is so scarce, the inferior kinds will pay for burning; and, as wood is cheap, there can be little risk trying the lime at some of the localities, both for agricultural purposes, and hydraulic cement.

4. PLASTER AND SALT.—The first has not been found at all, except in some few instances, in the tertiary clays.

The salt is frequently a mineral subordinate to the rocks of this series. It exists. Some of the waters issuing from these sandstones contain a small quantity of salt,—muriate of soda, or, more properly, chloride of sodium. I have obtained it in small crystals, by evaporation. The question of the existence of muriate of soda, in quantity, can only be settled by boring. There is one fact which seems to be unfavorable to its presence in sufficient quantities to become valuable. If the indications are to be relied upon, the rocks were deposited in deep water; and it appears that salt or brine springs are more commonly found in those which are formed under shallow water, and where the water itself evaporates under the sun sufficiently to crystallyze out of the liquid, and occasionally leaves a large area uncovered with water. But, however this may be, boring is justifiable; and, as numerous places are known where water furnishes salt, the expense attending the operation will not form a serious objection to such a project.

5. FREE STONE.—Dan and Deep Rivers both furnish, and may furnish, inexhaustible quantities of free stone, admirably adapted to all works of construction. The material is soft, when first removed from the beds, and hence is easily wrought into suitable forms; it hardens by exposure to the weather, and is therefore durable; its colors are bright, and the stone is therefore beautiful.

The taste and fashions of the times give a preference to building stones of this description; but durability has also had something to direct and settle public opinion. Chimneys which have been built of these stones have stood for fifty winters and summers; and yet their corners are as sharp as ever. Besides, it is not so subject to acquire mouldiness as granite. Granite, like some poor soils, encourages the growth of fungi, by giving them potash, or the alkalis; and hence, building, made of smoothly wrought granite, becomes dingy, especially if shaded. The expense of working free stone is much less than granite. Quarries may be opened on or near the navigable waters.

6. GRINDSTONE.—The kind of stone which is predominant is a sandstone. The grain is variable, from a coarse to a very fine grit. Among the grits, that kind which is suitable for grindstones is common. The series below the coal, as well as those above, furnish them. Among the grits, I have observed some very fine ones upon the south side of Deep River, not far from Mr. Campbell's. They appear to be adapted to the purpose of grinding finer cutlery. Experience, I believe, proves the value of these stones for the ordinary uses of the farmer, the grinding of axes, &c.

Very little attention, however, has been given to inquiry respecting the best beds. Should a market be opened, grindstones of the best quality can be obtained. Their color is both brown and gray. Their grit is very sharp, and the grades of hardness required for different purposes may be easily supplied.

7. MILLSTONES.—I am not sufficiently well informed, as to what state of perfection the millstones of Deep River may be brought. They are among the best stones for grinding corn. Whether art can make them best, or as good as the French burr stones, will be better determined by those acquainted with the manufactory of them, than myself. They are esteemed for corn, and this fact has given them credit and a market to almost any extent; and it will increase, provided means of cheap transport are provided: as they can be furnished much cheaper than French burr stone, and are equally good for some purposes.

§ 8. SHALE.—The slate of the coal series, being fragile, and easily decomposed, may be employed upon the soil, as a fertilizer. It is composed of alumina, silex, a little lime, phosphate of lime, and some potash.

Those layers, which abound in the *cypris*, and posidonia, are richest in phosphate of lime.

The composition adapts the use of it to sandy or loamy soils; and, though I do not venture to recommend their transportation far, yet, on the plantations, which have a poor soil, which are adjacent to this marl, it will pay well for hauling. It should be ground and sown freely, broad cast. These slates contain many hard oval bodies, which consist of silica, lime and phosphate of lime: the latter, in the proportion of more than one half. These long oval bodies are the excrements of fish, or lizards, which swarmed in the sea, in the days during the deposit of the system.

The recommendation is encouraged on the ground, that fertilizers are expensive in that region of country where this formation exists; and if it should be found useful, the country through which these shales pass, can be supplied, to any extent which is desirable.

It is rare, that a formation, which looks so unpromising on its first acquaintance, should turn out so rich in products, which will encourage industry and contribute so much to the advancement of wealth and prosperity.

The English new red sandstone, which is certainly closely allied to this formation, supports no less than nineteen large cities. It is true, that in that country, rock salt is one of the products of the red sandstone formation, which has been discovered here; but there is coal, which is still better, and which can promote the wealth of the Dan and Deep Rivers, to a far greater extent, than salt alone. The climate, and the health of the country, too, is in its favor. The navigable waters, or those susceptible of being made so; the value of the forests, in pines and oaks; the iron; all of which mark the Dan and Deep River, places these districts, in a position, equal to that of the country referred to, and if that can support and cherish the inhabitants of nineteen cities, certainly, this formation should give origin at least to four or five large and flourishing towns.

A careful survey of our own country, and others abroad, accompanied with an inquiry into the causes of the rise of cities and towns, will probably show, that those causes are mainly geological. It will show, that the products of the soil and the mine lie at the foundation of all the operations which have given rise to their establishment and subsequent prosperity. Pittsburg, in Pennsylvania, owes her origin to the iron and coal in her neighborhood. Rochester, in New York, owes her origin to the peculiar rocks there, whose constitution produces the Falls upon the Genessee, at this place, and those peculiar rocks give the surrounding country a wheat soil. Upon these facts, Rochester has become one of the most flourishing cities in the Union; and yet all these causes are geological. Deep River and the Dan have all these advantages and more.

REASONS WHY THE NEW RED SANDSTONES OF THIS COUNTRY DIFFER FROM THOSE OF EUROPE.

§ 109. The new red sandstone, in England, is underlaid by limestones, or calcareous rocks, to a greater or less extent. Some of them are magnesia; and hence, in the series, one of the members is strongly marked, and is known as the magnesia limestone. The origin and source of the materials appear to be entirely different; and hence, the lithological character of the series; and new red sandstone is quite different, at least in its subordinate parts.

In the United States, the materials are deficient in lime and magnesia; and on account of the presence of certain minerals diffused in the waters, and forming deposits, and thereby imparting a character to the whole sea,—these circumstances cannot fail to influence both animal and vegetable life; The sea-bottom will favor, or it will be unfriendly to the existence of certain species. To facts of this kind we may look for an explanation of certain modifications which are known to exist in the fossils of those rocks.

The marl slates resemble those of the Permian system. In Germany, they contain copper. Here, they are entirely destitute of copper. In other respects, they are quite similar. While it cannot be proved that rocks which contain the coal of Deep and Dan Rivers are Permian; still reasons are not wanting which favor this view: though the Richmond coal field is now regarded as belonging to the Oolite. I am, however, upon the whole, and on consideration of all the facts, inclined to adopt the opinion, that the whole series belongs to the upper new red sandstone. I am sure the great abundance of coal favors the view that this series should be regarded as Permian. So, also, the tooth of the Thecodontosaurus, or a saurian closely allied to it—but the most abundant fossil, the posidonia minuta, (*Goldf.*) favors more strongly the opinion I have adopted under the existing facts.

MISCELLANEOUS NOTICES OF MINERAL DEPOSITS AND VEINS.

Iron Ore.—In Nash County, I visited a deposit of Iron which had been worked, but now abandoned. It resembles the bog ores; should be classed with them; being simply a superficial deposit, of no great depth. This, in fact, is of no value. It is one of those formations which, I believe, has originated from ancient mineral springs, whose waters were charged with bicarbonate of iron; or an oxyde held in solution in a carbonated water When combinations of this character reach the surface, the carbonic acid escapes. When the iron is no longer soluble in water, it is precipitated upon the surface. There will then be found a deposit of oxyde of iron, intermixed with clay, sand, &c. Some of these deposits may contain sufficient iron to become valuable;—this will not. The extent should be determined by sounding with a slender bar of iron or steel, before expenditures are made.

Magnetic Ore, in Guilford County.—Magnetic ore of a fine quality, exists at Mr. C. Coffin's, ten miles from Greensboro'. It is free from sulphate of iron. It had not been examined, when I visited it, in a shaft. The surface ore presents a favorable indication of two or more veins of a fine quality.

Specular Ore, on the Plantation of Wm. Jones.—This ore is also a fine kind of this species; but its extent has not been determined by actual exploration.

Specular Ore on the Plantation of Mr. Glass. This location is six miles north of Evans's Mills. I regard this as the peroxyde, or the specular ore ; as it is un-magnetic, and gives a red streak. It is abundant, and, being in the vicinity of water power, it will come into use when the Deep River improvements are completed.

———o———

STITH'S COPPER MINE, IN GUILFORD.

§ 111. It has been known for a very long time, that the auriferous pyrites consisted in part of the sulphuret of iron, and, in part, of the sulphuret of copper. In extracting the gold from the sulphurets, the latter has been neglected and allowed to flow away in the washings. Lately, however, attempts have been made, not only to save the copper of the auriferous pyrites, but to work the veins exclusively for copper. Stith's mine had been worked for its gold for many years. It was profitable ; but its owner, Mr. Fentress, had given up the business of working it for gold, and it was lying useless to himself, when Mr. Stith proposed working the sulphuret for copper. Two shafts had been sunk upon the vein, at a distance of 316 feet; and, for some distance from each shaft, the ore had been removed and worked for gold. The vein runs N. 30 degrees E.; dip N. W. At the depth of almost 72 feet, the vein of pyrites is divided into two, [a flat vein, which dips about 5 degrees, and a vein dipping between 60 and 70 degrees.] The flat vein consists of a gangue of quartz, arranged somewhat in columns, and the vein of sulphuret, ranging in with from 4 to 12 inches: the whole width of the quartz and copper is from 2½ to 5 feet. This flat vein dips towards the steep dipping vein, and finally becomes incorporated with it, when it becomes the main and important vein of the mine.

The progress of the work becomes more and more favorable, and a fine vein of sulphuret of copper is likely to be disclosed, and, indeed, is so, by the present operations. The double sulphurets are changed to the single sulphurets, and it is found to yield from 32 to 40 per cent. of copper. The mine is valuable, and its success will operate favorably in producing a change in the working of the auriferous pyrites. The probability is, that many others, in which the copper has been lost, from ignorance of the value of the substance, will be worked so as to save the copper, or to work them as copper mines exclusively.

———o———

LIMESTONE.

§ 112. The great value and importance of limestone has created a demand for it, both as an article essential in construction, as well as in agriculture. In a very large part of North Carolina, this rock seems to be absent, and hence it has been difficult to supply lime sufficient only to meet the ordinary wants of the community. It has been always too expensive to warrant its employment for agriculture, and much of the loss in agricultural products may be attributed to the scarcity and expense of lime. Probably all the soils of this State will be benefitted by the application of lime. I have visited only the two well known localities of limestone in Stokes, the limestone belonging to Mr. Martin of ———, and Mr. Bolejack of Germanton. These beds of limestone belong to the pyro crystalline rocks. The stratification of Mr. Martin's beds is quite obscure, while that of Mr. Bolejack's is quite distinct. Both belong to the same kind of rocks.

The thickness of both exceeds forty feet, and lie between strata of coarse talcose slates—or talco-micaceous slate. Both beds make good lime. These beds may become in the hands of enterprising men both profitable to the owners and highly the useful to community. Mr. Bolejack's is located very conveniently for cheap mining, and wood being abundant and cheap, I have no doubt it may be furnished at 15 cents per bushel and perhaps 12½. At those prices the farmer can afford to use lime.

The beds seem to be in range with others crossing the State from N. E. to South West.

———o———

SOME OF THE GEOLOGICAL CHARACTERISTICS OF THE SLATES OF STOKES, SURRY, &c.

The predominant rock of these Counties is Talcose Slates with a variety which may be called talco-micaceous slate. The rock has the usual silvery lustre, and thin lamination, which is frequently undulating. The rock is generally covered with soil. The ridges and mountains are sharp and narrow, and present in out line a singular and picturesque appearance. This is especially the case with the Pilot mountain. From Germanton and other points, it presents the appearance of a high isolated rounded knob, bearing upon its summit a square tower. Seen from the residence of its owner, Mr. Guillam, it becomes a sharp ridge surmounted by two pinnacles—the eastern the greater of the two. The mountain sides are steep and precipitous. The pinnacles are bounded by perpendicular sides. The highest and most prominent one is ascended by means of ladders, and rises about 70 feet above the crest of the mountain.

These magnificent pinnacles have been formed by a very simple geological operation. The rocks were thrust up·wards in such a manner as to produce a decided curvature of the crest of the mountain, and so much of a curvature, as to produce a cross fracture of the strata between the pinnacles, which are 250 yards apart. The slow operation of at. mospheric agents have done the rest. These operations consisted in the disintegration of the softer slates, especially along the line of parture between the pinnacles. The undermined strata form the debris of the mountain sides. The harder strata of the pinnacles have withstood the action of the elements, and will stand and battle them for thousands of years to come. The strata of the pinnacles differ from each other. Some of the strata consist of pure granular quartz, especially those which form the pinnacles. These strata, however, should not be regarded as a sandstone, but simply a very quartzose variety of talcose slate. The Pilot and other mountains of the range belong to the first and most easterly of the Blue Ridge or Alleghanies ; but unlike other ridges, they are steepest on the eastern slope. The Pi·ot mountain is one of the greatest places in North Carolina. Nature has performed a work here, which seems to have been designed to give health and pleasure to those who have become debilitated or worn down under the burning and sultry atmosphere of the South. It is a pity, when so little is left to be done, to make the Pilot a place of great resort, nothing but a rough path way and a few ladders have yet been contributed to promote objects of so much importance. The geological structure of much of North Carolina is characterized by low anticlynal and synclynal axes. Some of the synclynal are deep and form troughs in which the coal fields lie. The axes are formed by normal dips, being equal on both sides of the rounded ridge.

CONCLUSION.

I have introduced a greater amount of elementary matter, perhaps, than is required in a simple Report, designed to give a statement of what has been done to carry out the plan of the survey. I have done this because many of the persons into whose hands this report will fall, wish something of the kind. Much of the elementary matter of the foregoing report has been published before, but I have proposed to make a direct application of these elements to the agriculture of the State.

The State of North Carolina might be divided into two great districts, the *Agricultural* and *Mining*—the former embraces those Counties which lie immediately upon the Atlantic slope, extending to the first fall of the rivers, where they enter the tertiary formation. The latter embraces all west of these falls. While the former, however, is eminently agricultural, the latter is both agricultural and mining.—Usually, a mining district is rough and comparatively unproductive; here, however, while mining gives, or is capable of giving, magnificent return, the agricultural is equally productive with other districts. The means of living are therefore cheap, and while a portion of its citizens are engaged in those pursuits which neither make a blade of grass, or potatoes grow, yet their labor always secures an abundance of bread and meat from the very surface beneath which the mineral wealth is drawn.

In pursuing the work up to the present time, I have scarcely touched upon the mining wealth of the State. The most I have attempted to do, is to to develop the value of the coal mines. The gold, copper, lead and iron mines, I propose to examine the ensuing year.

It is a remarkable fact, that, while lead and zinc are comparatively rare, gold and silver are abundant. I had occasion to notice a fact of like kind, in my Report of the

Geology of New York. In the Northern Counties of that State, iron is the great mining product ; it is accompanied with neither copper, lead, zinc or gold. I mean that it preponderates over every other metal. Iron occupies an important place in North Carolina ; and I may here say that the advantages for making bar iron of the best quality are very great. The ore in the first place is abundant and of an excellent quality ; and in the second place, wood for charcoal is equally abundant, and as the growth of trees is rapid, fuel will never fail if system is observed in its cutting and preservation of young timber. The resources of the forest in North Carolina are immense, notwithstanding a terrible disease has infested certain portions of it for some time past. The famous long leaf pine is a magnificent tree of the forest. It yields its turpentine and rosin in profusion—one of the great staples of the South ; its leaf makes an elegant hat, its cone an ornamental basket, its heart the most durable of posts, and its wood the cheerful fire and light, both of the kitchen and parlor. The great variety of Oaks and Walnut are no less important. The Tulip in beauty is rarely excelled, and the Magnolia among the trees of the forest is like a gigantic rose.

The water power is also immense. The improvements on Deep River and Cape Fear will furnish water for several Lowells. In fine, the elements of wealth and prosperity have been dealt out with a liberal hand, and its people have only to put forth their energy and enterprize, to stand with the first States in this repubiic.

DRIFT—DILUVIAL ACTION.

§ 116. In the Northern States and Canada, the surface of the country is overspread with a coating of soil stones, gravel boulders, etc., which are foreign to spots and places upon which they now rest. These materials have been transported from distant points, either North, or Northeast, from the spots we now find them, and, in many cases, more than one hundred miles from their parent beds. I wish merely to allude to this fact. It is a practical one; for, as the surface has not been disturbed, and as the disintegrations of rocks have gone on quietly, the debris remain in place. Hence, a mass of iron ore, or of copper, gold, etc., which lies upon the surface and in the debris, the parent bed or vein of each, will be found below, or at most, but a short distance from the spot; whereas, at the North, it is common to find a mass of iron ore which is one hundred miles from its bed or vein. In the latter instance, we know only the direction the mass has been transported. In North Carolina, we may always expect to find the ore in the immediate vicinity in which it is found, except in those cases where the loose mass has been removed by aqueous causes now in operation.

[As I was unable to incorporate the observations and remarks of Dr. McClenahan, one of my assistants, with my own, I deem it proper to give them a separate place in the report. They are, as will be seen, addressed to me in the form of a letter. They were made during my absence from the field, and while engaged in the laboratory :]

LETTER OF DR. McCLENAHAN.

Dear Sir :

After parting with you at Goldsborough, and arriving on the Coal Field, I commenced the survey of the underlying sandstone, at Captain Elias Bryan's, on the Deep River, one mile above Haywood. The dip at that point is South, 45 degrees East, at an angle of 20 degrees : the strike South, 45 degrees West. The sandstone and conglomerate are both properly exposed at this place, the sandstone resting immediately on the conglomerate. I commenced by running South, 45 degrees West, to Womble's : thence across the Haywood road, by Mrs. Gilmour's : thence by Mrs. Beddle's : thence due West, to Mathew Wicker's, (distant from the starting point, ten miles) : thence North, 35 degrees West, crossing the river, to Watson's, on the North side of the river : thence North, 70 degrees West, to Burns' Spring : thence due West, by J. Hasley's : thence South, 50 degrees West, by Richard Dowd's : thence South, 55 degrees West, by John Dowd's : thence South, 45 degrees West, crossing Indian Creek just above William Hays' : thence South, 50 degrees West, to Deep River, in Mrs. Street's plantation : thence South, 60 degrees West, crossing the river to the mouth of William Hancock's lane, in Moore county : thence South, 45 degrees West, to Sewel's quarry of conglomerate : thence by Davis' quarry : thence by Neil Dunlap's : thence by Allen McDaniel's : thence by Jesse Thomas', on Drowning Creek, in Montgomery county : thence South, 60 degrees West, by Calvin Rush's, on Mountain Creek : thence South, 45 degrees West, by David Harriss' : thence by Lucas' store : thence by John C. Chambers' : thence across Little River, two miles above Steel's bridge, in Richmond county : thence across Pe-Dee River, at the mouth of Brown's Creek, in Anson county . thence up the Northwest side of Brown's Creek, by the Carolina College : thence South, 60 degrees West, to the South Carolina line, in the Southeastern corner of Union county. I took cross sections, at nearly all the public roads which crossed the sandstone transversely, and found it varying in width, from six to fourteen miles. I frequently got the dip where the stone was well exposed, and it varies from 10 degrees to 60. I also made cross sections from six coal pits, out to the out crop of the underlying sandstone, and found it varying from one mile and three-fourths, to three miles : the greater the dip, the shorter the distance.

After running the line, to the South Carolina line, I returned to the starting point (Capt. E. Bryan's), and commenced running Northeast, across Deep and Haw Rivers, one mile Northwest of the town of Haywood : thence North, 30 degrees East, by Willam Crump's and William Bland's : thence by Neill Womble's, in whose field the conglomerate is well exposed : thence by Mrs. Amsled's, on New Hope Creek: thence across the creek, by William Clark's, Thos. Womble's, John Bland's, Causby Stone's, in whose plantation it again crosses the Creek : thence up the Northeast side of the Creek, but occasionally crossing and re-crossing, by Mooring's, by Herndon's old store, in Orange county : thence by Pratt's store, crossing the Central Railroad half a mile Northwest of the store : thence across Eno and Flat River, in Benehan's plantation ; after which, it could be but indistinctly traced. Although this is the direction of the great body of the stone, there is, occasionally, points which run off in various directions : one of the principal points which make off in this way, is one that continues up New Hope, to Morgan's creek, and up that creek to within two miles of Chapel Hill.

There is a formation of sandstone on Tau River. I saw it at Thos. Miller's plantation, six or seven miles Southwest of Oxford. I had understood that coal had been found there ;

but, when I examined the spot, which is in the river bank, I found it to be lignite. I have samples of it in Raleigh, and also of the micaceous sandstone in which it is embedded. My attention has been frequently called to the subject of lime, and I have been frequently told, that there was limestone on certain lands, which I was going to examine; but, as yet, I have not been enabled to discover lime in sufficient quantities to render it of much value, East of Germanton. I have seen small deposites of limestone in the upper stratum of what I have called the newer red sandstone. I found it at Mr. Fowler's, in Chatham, near Mooring's, and on the Hillsborough road, near Brassfield's, sixteen miles from Raleigh, and in Granville county, on the plantation of Mr. Wortham: it is in greater abundance at this point than at any I noticed. Mr. Wortham has hauled out on his farm a considerable quantity of it, and informed me that the land on which he spread it, produced much better. Lime in great abundance, and of excellent quality, is found stretching across the State, from Danbury, in Stokes county, to King's Mountain, in South Carolina. I saw it at Williams' kiln, on the Yadkin, at Poff's, ten miles above Salem, at Hoosertown, at Germanton, and at Martin's, near the Virginia line. I procured a piece near Germanton, at Mr. Bolejack's, which is an excellent marble, and receives a fine polish. The quantity of limestone at this point, appears to be inexhaustible, and of good quality; in fact, all the lime I saw at all the kilns appeared to be of good quality. I have procured samples of the stone from all the kilns, for your inspection. This section of the State abounds in iron ore of good quality. I have specimens from several places. Magnetic iron ore of good quality is found two miles West of the Pilot Mountain, on the lands of Mr. Guillam. I examined the place and saw it scattered over a large surface.

After examining the limestone, I commenced the survey of the coal field on the Dan River. I commenced at Germanton: the out crop of sandstone is near that place. The dip is Northwest, at an angle of 35 degrees, and the strike Northeast. I was able to trace the out crop of sandstone as far as

Madison, and have procured samples of the coal and slate, at various points; but, in consequence of high waters, I was unable to ascertain the thickness of the coal seam. The fossils are of the same kind we find on the Deep River, but the coal is anthracite. I should have continued the survey down to Leaksville, or as far as the coal continued in the State, but for high water.

The out crop of black shale, on the coal field, is in great abundance, and the direction of the seams can easily be traced from one end of the field to the other, with the appropriate fossils, in great abundance. I found but few points on the Deep River coal field, South of Deep River, where the shale could be easily traced. I found it on Drowning Creek, in Montgomery county, about one mile Northwest of the sandstone, which contains lignite: the dip at this point is not more than 10 degrees. I also found it at the Pe-Dee River, on the plantation of Mrs. McCloud.

After parting with you at Halifax, I visited the Northwestern corner of Edgecombe county, for the purpose of ascertaining the truth of what I had heard of a large skeleton which was embedded in Fishing Creek. I ascertained it to be the remains of an enormous whale, some of the vertebræ of which measured twenty-two inches in diameter. It had been so much mutilated, that I was deterred from attempting to disinter but a small portion of it. I learned from the gentleman, who owns the land in which it is embedded, that the largest portion of the bones had been taken away by various persons, some of whom lived at a great distance; and he also informed me that a large number of the bones had been washed away by the "freshets." I ascertained, by finding one or two vertebræ in place, that the animal had been deposited on his back, and as the water is not more than two or three feet above the vertebræ, which is just covered with marl and sand, I could readily account for the absence of all the ribs, by freshets, which swept them down the stream.

This animal is lying on a bed of marl, which is twelve or fifteen feet thick, and the silicious shelly limestone, which is found between the green sand and shell marl, is just above the remains: above that is a bed of yellow sand and shell marl, which is about seven or eight feet thick. Mr. Knight, the gentleman who owns the lands, told me that there was a portion of the head still embedded in the bank, and but for the rise which took place in the creek, while I was there, I should have procured it. The animal lay diagonally across the creek—the head in Edgecombe and the tail in Halifax, the creek being the line dividing the two counties. I picked up a good many of the bones, and requested Mr. Knight to take care of them for me, which he promised to do, and gave me the balance, if I could procure them. I procured some of the marl below the remains, and some of the upper bed, which is above it. I also procured a specimen of the shell rock, which is between the two beds. I have a piece of the jaw-bone in Raleigh, which I got out of the water near the spot where Mr. Knight told me the head was embedded in the bank.

After passing over the tertiary system, which continues, in the direction to Raleigh, about twenty-five miles above Nashville, I discovered the primary slates, talcose and micaceous, with a great many quartz veins running through them, showing strong indications of gold. The dip of these slates is to the South, 70 degrees East, at an angle varying from 25 degrees to 60 degrees; the strike South, 20 degrees West. After passing over this formation, I came to a formation of inferior granite, composed chiefly of feldspar and quartz, with a very small proportion of mica. This stone readily disintegrates when exposed to the frost, producing a coarse gravelly soil, which is an excellent land for corn, cotton and oats. This granite gradually increases until it reaches Raleigh, where it has a sufficient amount of mica to form a very good building stone. At Raleigh, the dip of the slate is changed from Southeast to Northwest, at angles varying from 25 degrees to 80 degrees; in fact, the dip near the Plumbago veins, four or five miles Northwest of Raleigh, is nearly perpendicular. The strike being South, 20 degrees West. I have procured specimens of this graphite for your inspection, from several points: some of it is of good quality, but the most of it that I saw was out crop, and, therefore, was full of dirt. I think these veins of graphite, by proper management, might be made immensely valuable. I should expect to find the mineral of much better quality, after going down to water scale. The stratum of Plumbago is of good size, from all appearances, but I was not able to measure it with accuracy, in consequence of the pits being filled with water.

I visited Mr. Johnson Busbee's, ten miles Southeast of Raleigh, where there are strong indications of marl. I found there the silicious shelly limestone, scattered over a large surface, the same which we have usually found between the green sand and shell marl. I have a piece for your examination. This point, I think, should be particnlarly examined. Marl, as far up the country as this, would be very valuable, in consequence of the scarcity of lime in that section of the State.

There is lignite in the Cape Fear River, about eighteen miles above Fayetteville, on Silver Run creek, and the sandstone, in which it is embedded, resembles that at Elizabeth, which we found between the two beds of marl. At this point is also found petrified wood. I think you would find this neighborhood an interesting one for examination, and the citizens are exceedingly anxious you should visit them, for that purpose.

I have samples of iron ore procured at various points in Cumberland county, which is all very silicious; probably too much so, to be of much value.

The above, I believe, constitutes all the information I have been able to procure during your absence. You will please examine the contents, make corrections of any mistake, and use it as you may think most advisable.

I remain your obedient servant,

S. McLENAHAN.

Professor Emmons,
State Geologist, N. C.

APPENDIX.

I have been obliged to refer to the different systems of rocks, in the foregoing report. I am induced, therefore, to furnish a tabular view of those systems, that the reader may be able, at a glance, to see the relations in which they stand to each other. I make three principal classes of rocks, which hold an equal rank. These three classes are subdivided. These subdivisions are based upon facts and phenomena, which are peculiar to each, and on characters which are not common to each division. The names of the principal classes are new, and are simply expressive of facts, upon which all geologists are agreed.

The three classes :

I. Pyrocrystalline—crystallized by the agency of fire. Primary of authors.
II. Pyroplastic—moulded by fire. Ancient and modern volcanic rock of authors.
III. Hydroplastic—moulded by water. Sediments of authors.

The first class is divided into two sections :

1. Unstratified pyrocrystalline, as granite, Hyperthene rock, pyrocrystalline limestone, sienite, magnetic iron ores.
2. Stratified pyrocrystalline gneiss, mica slate, talcose slate and hornblende steatite.

The second class is divided in two sections, also :

1. Modern pyroplastic rocks, lavas, tuffs, pumice and all the products of volcanoes, which are cooled in the air.
2. Ancient pyroplastic rocks, the ancient lavas, cooled under water, basalt, porphyry and green stone.

The third class is divide i into systems, most of which are admitted by geologists of the day.

The systems belonging to the class of hydroplastic rocks, the consolidated and loose sediments, aie exhibited in the following table :

I. Tertiary system :
1. Postpliocene.
2. Pliocene.
3. Miocene.
4. Eocene.

II. Cretaceous system :
1. Upper cretaceous, including the true chalk, with flints.
2. Lower cretaceous, including the green s und, iron sands, &c.

III. Wealden, unknown in the U. S.

IV. Oolite and Lias.

V. New red Sandstone or Trias :
1. Upper.
2. Middle.
3. Lower.

VI. Permian system.

VII. Carboniferous system.

VIII. Devonian system.

IX. Silurian system :
1. Upper.
2. Lower.

X. Taconic system.

The tenth is the oldest of the sediments, and is more closely allied to the primary or pyrocrystalline slates, limes, ores, etc. Any of the foregoing systems may rest on the primary, and any of the foregoing may be traversed by the unstratified pyrocrystalline rocks ; particularly granite, which is then said to be of the age of the deposit in which it is found. As any of the foregoing systems may rest upon the primary, so either may form the surface rocks over large areas.

GLOSSARY

OF SCIENTIFIC TERMS USED IN THIS REPORT.

Amorphous—shapeless, destitute of a regular form.
Arenaceous—sandy, composed of sand.
Argillaceous—composed of clay.

Basalt—a rock mostly homogeneous, of an igneous origin, and cooled under water.
Basin—a depression in the strata, of a circular form.
Belemnite—a fossil of acylindrical form, tapering rapidly to a point, and at one end or the other it has a conical cavity: it is the back-bone of an extinct animal, allied to the cuttle fish.
Bitumen—a combustible substance, combined with coal.
Breccia—a compound rock, consisting of angular fragments.

Calcareous—bearing or containing lime.
Calcedony—a compact variety of quartz, of a milky whiteness.
Carbon—the element of charcoal.

Carbonate of Lime—a compound of carbonic acid and lime.

Carboniferous—coal bearing: a term applied to a system of rocks which bear coal.

Cetacea—an order of animals, of which the whale is the type.

Chert—a variety of amorphous quartz, much like flint.

Concretion—a union of particles, forming rounded and oval bodies.

Conformable—a term applied to strata, which lie parallel with each other.

Conglomerate—A rock composed of rounded pebbles, formed under water.

Coniferæ—trees which bear cones, with naked seeds, as pines and the fir.

Cretaceous—belonging to chalk: the name of the system to which common chalk belongs.

Crustacea—an order of animals which are provided with a crust or external integument similar to the lobster and crab.

Dikes, or Dykes—a vein of rock or stony matter, which has been injected into a fissure, while in a melted state.

Diluvium—a term which was applied to a stratum, which was supposed to have been spread over the earth by the deluge.

Dip—strata, when inclined to the horizon, are said to dip.

Eocene—dawn of the present: a term applied to the oldest of the tertiary deposits.

Escarpment—the steep side of a hill.

Estuary—the mouth of a river, which is occupied, in part by fresh, in part by salt, water, or by brackish water.

Faults—the dislocation of strata, by which one side is elevated above the other.

Fauna—the aggregate of the animals which inhabit certain districts.

Formation—a series or group of rocks, which belong to one period.

Fossils—the remains of animals and plants entombed in rocks.

Gypsum—a rock or mineral, composed of oil of vitriol and lime.

In condescensu—minerals or rocks, in a state of fusion.

Laminated—a mineral or rock, composed of thin plates.

Lias—supposed to be derived frem layers: a system between the oolite and new red sandstone.

Lignite—wood carbonized or changed partly into coal.

Lithological—denotes the stony characters of a mass.

Littoral—belonging to the shore.

Loam—a mixture of sand, clay and vegetable matter.

Lycopodites—a fossil plant, allied to club masses or ground pines.

Mammalia—animals which furnish glands for the secretion of milk.

Mammoth—an extinct thick-skinned animal, allied to the elephant.

Marl—a mixture of lime and clay.

Mastodon—see mammoth.

Miocene—the middle deposits belonging to the tertiary.

Molusca—an order of animals generally covered with shells, as the oyster and clam.

Nodule—a rounded mass.

Out crop—the appearance of the edges of rocks at the surface.

Oxygen—a gaseous body, which is essential : it changes the blood from a black to a scarlet color, in respiration; combines, with metals, and forms a class of bodies call. ed oxides, etc.

Pachydermata—an order of animals with thick skins, as the elephant, hog, tapir, horse, camel.

Palæonlology—the science which treats of extinct animals and plants.

Porphyry—an igneous rock, cooled beneath water, and which contains irregular pieces of feldspar.

Pyrites—sulphur and iron in combination.

Rodentia—an order of snimals, supplied with front cutting teeth, similar to the squirrel and rabbit; gnawer as the rat.

Ruminants, Ruminantia—an order of animals which chew the cud, as cow, sheep, deer.

Saurian—a lizard-like animal.

Schist—a rock made of three parallel layers.

Sediments, Sedimentary—mud, sand, etc., deposited under water.

Septaria nodules—composed of clay, lime, etc., divided into parts or partitions of crystalline matter.

Shale—indurated clay.

Silex—silica, flint.

Stratified—divided into layers.

Strike—the line of the bearing of rocks which lies at a right angle to their dip. The ridge-pole of a house shows the strike : the inclination of the roof, the dip; and it forms, in this illustration, an anticlynal axis.

Syenite—a variety of granite, in which hornblende takes place of mica.

Synclynal Axis—the reverse of anticlynal, when the strata, on two sides, plunge towards each other, or to a line below their out crop.

Trap—volcanic rocks.

Veins—Fissures filled with mineral matter, differing from the rock in which the fissure has been found.

Unconformable Strata—reposing upon the edges of strata or when the layers are not parallel to each other.

ERRATA.

In section 30, instead of reading "*Indian Corn,*" read *Cotton.*

In section 27, instead of $65 per acre, read $15 per acre; and instead of "two laborers," read "twelve laborers."

In section 28, fourth line from the bottom of the paragraph, for manures read *measures*. In same section, third line from the bottom of the paragraph, for prepared read *purchased.*

[The PUBLIC PRINTER thinks it probable that some typographical inaccuracies occur in the foregoing report. The unavoidable absence of the author, in the prosecution of his labors, devolved upon the publisher the duty of revising the proof-sheets. *His* want of familiarity with most of the technical terms employed, renders it probable that errors exist, so far as those terms are concerned.]